New Security Challenges

Series Editor
George Christou
University of Warwick
Coventry, UK

AF738331

The last decade has demonstrated that threats to security vary greatly in their causes and manifestations and that they invite interest and demand responses from the social sciences, civil society, and a very broad policy community. In the past, the avoidance of war was the primary objective, but with the end of the Cold War the retention of military defence as the centrepiece of international security agenda became untenable. There has been, therefore, a significant shift in emphasis away from traditional approaches to security to a new agenda that talks of the softer side of security, in terms of human security, economic security, and environmental security. The topical New Security Challenges series reflects this pressing political and research agenda.

* * *

For an informal discussion for a book in the series, please contact the series editor George Christou (G.Christou@warwick.ac.uk), or Palgrave editor Alina Yurova (alina.yurova@palgrave-usa.com).
This book series is indexed by Scopus.

Gertjan Boulet
Michael Reiterer • Ramon Pacheco Pardo
Editors

Cybersecurity Policy in the EU and South Korea from Consultation to Action

Theoretical and Comparative Perspectives

palgrave
macmillan

Editors
Gertjan Boulet
Korea University
Seoul, South Korea

Vrije Universiteit Brussel (VUB)
Brussels, Belgium

Ramon Pacheco Pardo
King's College
London, UK

Michael Reiterer
Centre for Security, Diplomacy & Strategy (CSDS)
Brussels School of Governance (BSoG-VUB)
Brussels, Belgium

ISSN 2731-0329 ISSN 2731-0337 (electronic)
New Security Challenges
ISBN 978-3-031-08386-0 ISBN 978-3-031-08384-6 (eBook)
https://doi.org/10.1007/978-3-031-08384-6

© The Editor(s) (if applicable) and The Author(s), under exclusive licence to Springer Nature Switzerland AG 2022
This work is subject to copyright. All rights are solely and exclusively licensed by the Publisher, whether the whole or part of the material is concerned, specifically the rights of translation, reprinting, reuse of illustrations, recitation, broadcasting, reproduction on microfilms or in any other physical way, and transmission or information storage and retrieval, electronic adaptation, computer software, or by similar or dissimilar methodology now known or hereafter developed.
The use of general descriptive names, registered names, trademarks, service marks, etc. in this publication does not imply, even in the absence of a specific statement, that such names are exempt from the relevant protective laws and regulations and therefore free for general use.
The publisher, the authors, and the editors are safe to assume that the advice and information in this book are believed to be true and accurate at the date of publication. Neither the publisher nor the authors or the editors give a warranty, expressed or implied, with respect to the material contained herein or for any errors or omissions that may have been made. The publisher remains neutral with regard to jurisdictional claims in published maps and institutional affiliations.

This Palgrave Macmillan imprint is published by the registered company Springer Nature Switzerland AG.
The registered company address is: Gewerbestrasse 11, 6330 Cham, Switzerland

Foreword

When invited to offer some thoughts on the perspectives for cybersecurity cooperation between the European Union and South Korea, I was reminded of the EU-Korea summit in 2013 in Brussels, where I had with my Korean counterpart the first substantial discussion of working together on cyber issues. We welcomed the launch of bilateral cyber policy consultations, and importantly, we agreed to continue joint efforts to maintain a safe, open and secure cyberspace. Convinced of the necessity to promote economic and social development by maximising the positive role of ICT and the Internet, we were also mindful of the projected important role of cyberspace in fostering and protecting on-line the core values of human dignity, freedom, democracy, equality, the rule of law and respect for human rights.

This volume shows that these seeds have borne fruit, even if the tree they rest on still needs significant care and engagement if it is to continue to blossom. As is often the case in diplomacy, moving from an exchange of information and best practices to action remains a challenge. This volume shows points of direction by identifying common goals and implementation tools: the means to connect people, to further invest in both commercial applications and a robust international governance framework that support democracy based on law, norms, rules and values.

Making good use of the internet and the social media has grown in importance and gained a further boost through covid-19, which prevented travel to a large degree. The Internet became the great connector. The cyber world gained even more importance, and in parallel the need to address cybersecurity grew further.

Beyond personal and commercial use, the military dimension of the Internet is growing. At the beginning of the war of aggression by Russia against Ukraine, cyber attacks sought to destabilise the country. My colleague, the former president of the European Commission, Jean-Claude Juncker, warned in his 2017 State of the Union address: "Cyber-attacks can be more dangerous to the stability of democracies than guns and tanks". Now we witness both—cyber attacks and tanks combined in hybrid warfare. In response to cyber attacks, armed forces added cyber attack and cyber defence units. The protection of the ever-growing critical infrastructure has become an essential aspect of security. Having been the victim of a cyber attack in my capacity as the president of the European Council in 2012, I know first-hand the importance of securing networks and accounts.

South Korea is one of the most connected societies; it invests in emerging technologies and related research, and leads in the production of semiconductors. South Korea has already successfully cooperated with European partners in the Horizon 2020 research programme on 5G technology, which is crucial for communication but also security. This intensification of cooperation between the EU and South Korea fits very well into the overall EU policy designed to enhance cooperation with like-minded partners in the Indo-Pacific region on strategic goals. The EU has recently intensified its engagement with Asia and the Indo-Pacific through programmes focusing on connectivity: the Connectivity Strategy with Asia, the Indo-Pacific Strategy and the Global Gateway. A common feature of all these programmes is to connect people and not only building physical infrastructure. Enhancing this cooperation and developing 6G together would add a highly political but also practical dimension to the strategic partnership of the EU and South Korea.

This also applies to strengthening cooperation in repelling cyber attacks and using countermeasures. South Korea's consideration of its bilateral relations with some countries means that it prefers to join multilateral responses and sanctions regimes rather than take its own unilateral measures. In this respect and our common interest, the EU could help to foster cooperation with neighbouring countries like Japan—I launched the "EU-Japan Cyber Dialogue" with Prime Minister Abe in 2014. In order to enhance cybersecurity a global networked approach is necessary to which a robust cyber diplomacy by the EU would contribute and to which South Korea could be invited to join.

Cyberspace has gained multidimensional importance in areas like war and peace, preserving and threatening democratic systems, providing a medium for real and fake news, respecting or even defending human rights. Mark Leonard alerted us to the danger of inaction in his recent book *The Age of Unpeace, How Connectivity Causes Conflict*: the Internet, the ultimate unifier of the global village, should not be allowed to become the perfect weapon to divide the world.

In short, this is a timely and well-curated volume of work by European and Asian experts, published at a time when the new administration of President Yoon Suk-yeol is forming its policies. I hope the new Administration in Seoul will draw some inspiration from this volume and continue building a stronger bilateral relationship between the EU and South Korea in cybersecurity and beyond.

President Emeritus European Council — Herman Van Rompuy
Minister of State

Foreword

This volume marks a signature publication in a continued process of cooperation with editor Gertjan Boulet: hosting him as a visiting researcher at Korea University in 2016—initially for a period of six months—planted the seeds for long-term cooperation to date. This volume in fact has its roots in an inspiring guest lecture of co-editor Michael Reiterer during one of Gertjan's courses at Korea University. As the editors say in the Introduction, the guest lecture on EU foreign policy in cyberspace led to a series of workshops bringing together Korean and EU experts from diverse fields, which we attended in a spirit of strengthening cyber cooperation between the EU and South Korea.

Digital technology has become more important than ever in various areas like smart cities, autonomous vehicles and artificial intelligence (AI). However, as being at the edge of emerging technologies grants states a competitive advantage, cybersecurity has become a matter of national survival; cyber attacks may cripple the infrastructure of a country, and thus we talk of the need to protect critical infrastructure. As there are no limits in time and space for communication and interaction between countries, international cooperation is the only way to build competitive digital economies throughout the world and to achieve efficient and practical cybersecurity. As various chapters in this volume show, the EU and South Korea already cooperate closely as strategic partners in a wide range of areas covering economic, trade, political, diplomatic and security sectors, and with cybersecurity as a prime area for joint actions. As part of the EU's Connectivity, Indo-Pacific and Global Gateway strategies, bilateral cooperation with South Korea should further advance, and the existing strategic partnership should be further developed in view of building highly efficient digital capabilities in the region and addressing challenges posed by digital transformation, including social and ethical issues.

The EU's growing weight on the digital economy and leading role in the regulation of data and AI has inspired South Korea's digital strategy, with the GDPR and the necessary adequacy assessment of Korean regulations as a representative example of the EU's influence. However, more work has to be done to achieve a truly global, open, free and secure cyberspace, based on regulations that are proportionate and strike a balance between government control and freedom.

As cyber threats become more and more sophisticated and complex, defence becomes difficult with existing security tools. When starting the illegal invasion of Ukraine in 2022, Russia conducted a series of cyber attacks on critical infrastructure in Ukraine and disseminated fake news to deceive Ukrainian citizens and the international media. Volunteer hackers and tech companies worldwide started to defend Ukraine and disrupt Russia's digital infrastructure through cyber attacks, actions motivated by the conviction that international peace must be maintained and that history should not repeat itself. Such unexpected consequences of the digital transformation can be expected to intensify in the future and should motivate the international community to develop practical measures to collectively resolve global cyber problems and serve as a catalyst to strengthening international voluntary measures even without any agreements on the global level. Strategically, the lesson to draw from Russia's invasion is to expand South Korea's relationships with NATO and EU in a multilateral perspective, also given the continued cyber attacks from North Korea. A wide range of sanctions by numerous countries have been imposed on Russia and Russian politicians after the aggression against Ukraine, organizations have been denouncing Russia's illegal acts, the UN General Assembly reprimanded Russia which was also excluded from the Council of Europe on March 16, 2022. On the same day, the International Court of Justice ordered Russia to suspend its military operation in the territory of Ukraine. These internationally concerted actions suggest that international cooperation can prevent and preclude potential or existing threats in any space including cyberspace. To avoid escalation, international cooperation among countries should be sophisticated and coherent. Cooperation between the EU and South Korea should move in this direction.

This volume is the first attempt to analyse bilateral cyber cooperation of two important global players in a comprehensive manner, emphasizing the need for EU-South Korea international cyber cooperation with concrete policy directions. Distinguished professionals in international law, diplomacy, international and cybersecurity have cooperated effectively drawing from each other's experience. The main focus of this volume is not on theoretical analysis but on the formulation of policy proposals in all relevant sectors including cyber diplomacy, data protection, cybersecurity, emerging technology regulation, cyber defence, cyber sanctions, arms control, digital investigation and international cyber governance. This approach ensures that readers and policymakers gain new perspectives and insights for devising future digital strategies and cyber policies.

We would like to thank the editors for the initiative and will build on it when deepening our cooperation with the EU.

School of Cybersecurity
Korea University, South Korea

Jong-In Lim
Hyoung-Joong Kim
Hun-Yeong Kwon

Contents

Introduction 1
Gertjan Boulet, Michael Reiterer, and Ramon Pacheco Pardo

Part I Rationale for Cybersecurity Cooperation Between the EU and South Korea 15

EU Cyber Diplomacy: Value- and Interest-Driven Foreign Policy with New Focus on the Indo-Pacific 17
Michael Reiterer

EU-South Korea Cooperation on Cybersecurity, Data Protection and Emerging Technologies 41
George Christou and Ji Soo Lee

Cyber Offence Dominance, Regional Dynamics, and Middle Power–led International Cooperation 67
Mason Richey

Developing a Collective Retorsion Framework Against Malicious Cyber Operations: Opportunities and Steps for EU-South Korea Cybersecurity Cooperation 99
Joohui Park

Part II Comparisons 117

Cyber Arms Control on the Korean Peninsula: Challenges and Opportunities with a View to North Korea's Asymmetry Strategy 119
Sangho Lee and Maximilian Ernst

Arms Control in the Cyber Domain: A European Approach to Mitigate Digital Threats 137
Michael Zinkanell

South Korea's Cyber Defence and Digital Cooperation with the EU 155
Kyu-dok Hong and Seong-jong Song

Building Cyber Resilience: The Defensive Shield for the EU 179
Anna-Maria Osula

Enhancing Cooperation Between South Korea and the EU in the Fight Against Cybercrime 197
Gibum Kim

The EU and Access to Electronic Evidence: Privatisation of Law Enforcement? 215
Tatiana Tropina

Part III International Order and Cross-Cutting Issues 233

Recent Trends in UN Cybersecurity Governance and South Korea-EU Cooperation 235
JoonKoo Yoo

The International Cybersecurity Cooperation Dilemma and Implications for EU-South Korea Relations 251
Cuihong Cai

Conclusion: From Words to Deeds 269
Michael Reiterer, Ramon Pacheco Pardo, and Gertjan Boulet

Index 285

Notes on Contributors

Gertjan Boulet is a lecturer and researcher in International Cybersecurity Law and Policy at Korea University, South Korea, and Vrije Universiteit Brussel (VUB), Belgium. He has written for various scientific journals and media, and as a frequent guest speaker addressed the Korean National Assembly, the Ministry of National Defense, and the Asia-Pacific Economic Cooperation (APEC)'s Asia Pacific Privacy Authorities (APPA) forum. Boulet has advised Belgium's Deputy Prime Minister and Minister of Telecommunication, the European External Action Service, and the Belgian Data Protection Authority.

Cuihong Cai is Professor of International Relations at the Center for American Studies, Fudan University, China. She worked for the Foreign Affairs Office of Fudan University (1996–2001) and was a visiting scholar at the Georgia Institute of Technology, USA, in 2002 and at the University of California, Berkeley, USA, in 2007, as well as an invited fellow in the 2007 program on US National Security sponsored by the US State Department. She is now a member of the Shanghai Association of American Studies. Cai is the author of *Cyberpolitics in U.S.-China Relations* (English version 2021; Chinese version 2019), *Political Development in the Cyber Age* (2015), *U.S. National Information Security Strategy* (2009), and *Internet and International Politics* (2003), as well as dozens of articles and papers on cyberpolitics, cybersecurity strategy, cyberspace governance, and Sino-US relations. Cai received her Ph.D. in International Relations from Fudan University in 2002.

George Christou is Professor of European Politics and Security at the University of Warwick, UK. Christou is the editor of the new security challenges series at Palgrave. His main research interests include the EU's role in conflict resolution/peacebuilding, European neighborhood policy, security governance, the EU as an actor in Internet governance, the political economy of European and global Internet governance, EU cybersecurity policy, and the politics of cybersecurity. Christou holds a Ph.D. in Political Science from the University of Sheffield, UK.

Maximilian Ernst is a Ph.D. researcher at the Centre for Security, Diplomacy and Strategy (CSDS) of the Brussels School of Governance, Vrije Universiteit Brussel (VUB), Belgium. His Ph.D. project is embedded within the Korea Foundation (KF)-VUB Korea Chair and focuses on the intercorrelation between political and economic cooperation among Northeast Asian states and the implications for European Common Security and Defence Policy (CSDP) towards the region. He worked as an investment consultant for the Korea Trade-Investment Promotion Agency (KOTRA) in Hamburg and as a political affairs trainee at the European Delegation to the ROK. Ernst is the managing editor of the *Global Politics Review* and the *Journal of International Studies*, and obtained his M.A. in International Security and Foreign Policy from Yonsei University, ROK.

Kyu-dok Hong is a professor in the Department of Political Science and International Relations at Sookmyung Women's University, ROK, since 1993. He served as the deputy minister for Defense Reform at the Ministry of National Defense, ROK, and as president of the International Policy Studies Institute of Korea (IpsiKor) ROK, and the Council for Security Cooperation in the Asia Pacific-Korea (CSCAP-Korea). He is also a member of the Seoul Forum for International Affairs. Hong received his Ph.D. from the University of South Carolina, USA, in 1991.

Gibum Kim is an associate professor in the Department of Forensics at Sungkyunkwan University, ROK. His research interests include cybercrime investigations, digital forensics, cybersecurity policy, and international development cooperation. He has worked as a police officer for a period of 23 years and had an Honorary Retirement as a senior superintendent in 2020. He worked as a professor in the Department of Police Science and as director of the International Cybercrime Research Center at the Korean National Police University, ROK. Kim received his Ph.D. from Korea University, ROK.

Sangho Lee is Professor of Politics and Diplomacy at Daejeon University, ROK. He previously served as an assistant administrator of the ROK Navy Development Committee and as chairman of the Korea Cyber Peace and Security Forum. He is the Vice President of The Korea Association of National Intelligence Studies. Lee was a researcher in the Department of Security Strategy Studies at the Sejong Institute, ROK; at the Korea Research Institute for Strategy (KRIS); and at the Centre for Defence Studies, King's College London, UK. Previously, he also served as an adjunct professor at Gyeonggi University and Hannam University, ROK. Lee received his Ph.D. in War Studies from King's College London.

Ji Soo Lee is a lecturer in the area of international trade/business law at the Hankuk University of Foreign Studies, ROK. She taught at highly reputed universities in South Korea (e.g., the Graduate School of International Studies at Yonsei University) and worked for various organizations in South Korea including government agencies (e.g., Ministry of Foreign Affairs), law firms, and research centers (e.g., Institute of Foreign Affairs and National Security). She also participated in the Free Trade Agreement negotiations and WTO dispute settlement procedures. Her ongoing research focuses on the OECD Guidelines for Multinational Enterprises and digital trade. Lee received her S.J.D. (Doctor of Juridical Science) in Law from the University of Wisconsin Law School, USA, in 2006.

Anna-Maria Osula Anna-Maria Osula is an Advisor for Cyber Diplomacy at the Ministry of Foreign Affairs of Estonia, a senior researcher at the Tallinn University of Technology, Estonia, and a research fellow of Masaryk University, Czechia. Previously, she served as Global Digital Governance Fellow at Stanford University, US, as a senior policy officer in Guardtime, Estonia, and as a legal researcher at the NATO Cooperative Cyber Defence Centre of Excellence, Estonia, where her research areas included national cyber security strategies, international organizations, international criminal cooperation, and law. During 2013–2015, she was the lead for the NATO Cooperative Cyber Defence Centre of Excellence (CCCDOE) 'Cyber Norms' project and the co-editor of the book *International Cyber Norms: Legal, Policy and Industry Perspectives.* Osula obtained her PhD in Law from the University of Tartu, Estonia.[1]

[1]Any views or opinions expressed in this article are personal and do not represent those of institutions or organisations that the author is associated with in her professional capacity.

Ramon Pacheco Pardo is the Korea Foundation-Vrije Universiteit Brussel (KF-VUB) Korea Chair at the Brussels School of Governance (BSoG-VUB), Belgium; Professor of International Relations at King's College London, UK; adjunct fellow (non-resident) with the Korea Chair at the Center for Strategic and International Studies (CSIS), USA; and non-resident fellow with the Sejong Institute, South Korea. Pardo has testified before the European Parliament and advised the OECD, the European External Action Service, the South Korean Ministry of Foreign Affairs, and the United Kingdom's Cabinet and Foreign & Commonwealth offices. He is a frequent media commentator on North East Asian affairs and EU-East Asia relations. His research focuses on East Asia's international relations and political economy (especially the Korean Peninsula), EU-East Asia relations (especially South Korea, North Korea, and China), and regional governance in East Asia. Pardo holds a Ph.D. in International Relations from the London School of Economics and Political Science (LSE).

Joohui Park is a senior researcher at Korea University Cyber Law Centre, ROK. She is a Lecturer in International Trade Law at Korea University, a member of International Cyber Law Studies, ROK, and a board member of Korea International Law Review. She has been participating in several governmental projects commissioned by the Ministry of Foreign Affairs, Ministry of Trade, Industry & Energy, and so on. Park recently finished a research fellowship sponsored by the Korea Society of International Economic Law with a paper on "Governing Cybersecurity Measures in the Digital Trade Agreements." Park received a Ph.D. in International Law from Korea University, with a dissertation on the "Status of Data under the Law of Armed Conflicts."

Michael Reiterer, former Ambassador of the European Union (Republic of Korea; Switzerland, Liechtenstein), is Distinguished Professor at the Centre for Security, Diplomacy & Strategy (CSDS), Brussels School of Governance (BSoG-VUB), Belgium, and Adjunct Professor of International Politics at the University of Innsbruck, Austria; Webster University, Vienna, Austria; Danube University Krems, Austria; and Libera Università Internazionale degli Studi Sociali "Guido Carli" (LUISS), Italy. In Japan he was a visiting professor at Keio, Ritsumeikan, and Kobe Universities. He is also an associate fellow at the Geneva Centre for Security Policy (GCSP) and the Vienna Institute for International Economic Studies (WIIW), and a senior advisor at the Center for Asia

Pacific Strategy (CAPS), Washington D.C., and STEAR (Student Think Tank for Europe-Asia Relations), Singapore. His ongoing research focuses on EU foreign policy, EU-Asia relations in particular with Japan and Korea, inter-regionalism, security issues, new forms of diplomacy, and human rights. He is also the editor of the CSDS Policy Briefs. Reiterer holds a Ph.D. in International Politics from the University of Innsbruck (AT), focusing on EU-Asia relations.

Mason Richey is Professor of International Politics at Hankuk University of Foreign Studies, ROK; senior contributor at the Asia Society, ROK; and co-author of the tri-annual US-Korea relations analysis chapter in Pacific Forum's *Comparative Connections.* Formerly he was POSCO Research Fellow at the East-West Institute (HI) and Deutscher Akademischer Austauschdienst (DAAD) Fellow at the University of Potsdam, Germany. His research articles on US and European foreign/security/defense policy have appeared (inter alia) in *Pacific Review*, *Global Governance*, and *Foreign Policy Analysis.* Most recently he is the co-editor of the volume *The Future of the Korean Peninsula: 2032 and Beyond* (2021). Richey received his Ph.D. from Binghamton University, New York, USA.

Seong-jong Song is a professor at Daejeon University, ROK, and director of the U.S. Affairs Center at the Korea Research Institute for National Strategy (KRINS), ROK. He was Disarmament Coordinator and Defense Attaché for the Permanent Mission of ROK to Geneva, Switzerland; Director of the U.S. Affairs Division at the ROK's Ministry of Defense; and counselor in charge of U.N. Peacekeeping Operations for the Permanent Mission of ROK to the UN. He retired from the ROK Army after more than three decades of military career, with a rank of Colonel in 2016. Song's major areas of interest include cybersecurity, ROK-US relations, US-China relations, civil-military relations, security policy, and military strategy of the USA, among others. He has written more than twenty peer-reviewed academic articles in major academic journals, including the *Journal of National Defense Studies*, *Journal of North-East Asian Studies*, and *Military History*, among others. Song received a Ph.D. in International Relations from the University of Missouri, Columbia, USA, in 1996.

Tatiana Tropina is Assistant Professor of Cybersecurity Governance at the Institute of Security and Global Affairs, Leiden University, Netherlands, and an associate fellow with The Hague Program for Cyber Norms, Netherlands. Previously, she worked as a senior researcher at the Max

Planck Institute for the Study of Crime, Security and Law, Germany. She has more than 15 years of experience in academic research, policy, and advocacy in the field of cybercrime, cybersecurity, ICT regulation, and Internet governance, including volunteer involvement in policymaking and Internet governance processes at Internet Corporation for Assigned Names and Numbers (ICANN). She was head of the legal department of several telecommunication companies (2003–2008) and worked on several international projects including a cybercrime study for the Global Symposium of Regulators (ITU, 2010), a UNODC (United Nations Office on Drugs and Crime) Comprehensive Cybercrime Study (2012–2013), the World Bank's World Development Report 2016, and a project with the German Federal Criminal Police Office on mutual legal assistance on interception of electronic communications in the EU (2015–2018). Tropina received a Ph.D. in Criminal Law and Criminology from the Far Eastern Federal University, Russia.

JoonKoo Yoo is a research professor at the Center for International Law, Korea National Diplomatic Academy (KNDA) of the Ministry of Foreign Affairs, ROK, and an invited professor at the Sungkyunkwan University Law School, ROK. He has been an advisory committee member at the ROK's Ministry of Foreign Affairs; Ministry of Defense; Ministry of Science, Information & Technology; and Ministry of Trade, Industry & Energy. He has also served as a deputy director of the Presidential Committee for the G20 Seoul Summit (2010–2011), and as a chief legal advisor for the UN Group of Governmental Experts (UNGGE) and the Open-Ended Working Group (OEWG) on Developments in the Field of Information and Telecommunications in the Context of International Security, and for the UN Group of Governmental Experts on Lethal Autonomous Weapon Systems (LAWS). Before teaching, Yoo was a legal advisor, specializing in international trade and defense acquisition with Aitken, Berlin & Brooman in Washington D.C., USA. He currently specializes in science & technology governance and laws, focusing on cyberspace and outer space, and artificial intelligence (AI). Yoo obtained a Ph.D. in Law from Sungkyunkwan University Law School, ROK.

Michael Zinkanell is deputy director of the Austrian Institute for European and Security Policy, Austria. His research focus lies on analyzing new dimensions of hybrid threats and emphasizes the political implications of disinformation and cyberattacks. Zinkanell is also involved in the European Cyber Young Leaders initiative, a network of European professionals with various backgrounds in cybersecurity. He obtained his M.A. in International Development from the University of Vienna, Austria. During the research for his master's thesis on Japanese development initiatives in Afghanistan and Iraq, Zinkanell worked at the Afghan Embassy in Tokyo.

List of Tables

EU-South Korea Cooperation on Cybersecurity, Data Protection and Emerging Technologies
Table 1 EU-South Korea: Key Issues and Further Cooperation 61

Cyber Arms Control on the Korean Peninsula: Challenges and Opportunities with a View to North Korea's Asymmetry Strategy
Table 1 Selected arms control agreements and their applicability to cyberspace 130

Introduction

Gertjan Boulet, Michael Reiterer,
and Ramon Pacheco Pardo

1 Areas for Cyber Cooperation Between the European Union and South Korea

Digital economy, cyber threats and rules-based cooperation are the three areas of cyber cooperation between the EU and South Korea, which are also reflected in the 2021 EU Indo-Pacific Strategy. Growing

G. Boulet (✉)
Korea University, Seoul, South Korea

Vrije Universiteit Brussel (VUB), Brussels, Belgium
e-mail: gertjan.boulet@vub.be

M. Reiterer
Centre for Security, Diplomacy & Strategy (CSDS), Brussels School of Governance (BSoG-VUB), Brussels, Belgium
e-mail: michael.reiterer@vub.be

R. Pacheco Pardo
King's College, London, UK
e-mail: Ramon.Pacheco.Pardo@vub.be

© The Author(s), under exclusive license to Springer Nature Switzerland AG 2022

G. Boulet et al. (eds.), *Cybersecurity Policy in the EU and South Korea from Consultation to Action*, New Security Challenges,
https://doi.org/10.1007/978-3-031-08384-6_1

digitalisation increases cyber threats worldwide, exacerbated by the strategic competition between the US and China. Mindful to preserve some autonomy in this technological and geopolitical competition, the EU and South Korea share common incentives to cooperate particularly in these three areas.

First, on **digital economy**: because of digitalisation, trade conflicts between the US and China have expanded to conflicts over digital technology. Prohibiting export of digital technology and blocking tech acquisitions in investment control are tools to limit access to digital technology by other countries. In this respect, 'chip wars', or 'wars' over semiconductors or microchips (an essential component of electronic devices such as mobile phones and computers and less visibly to the public in the automotive industry), are of particular importance. Microchips are key in building digital connectivity worldwide. They empower states with dominant digital service providers to weaponise trade in digital products and services, or to instrumentalise digital supply chains for political and military ties and warfare. Thus, digital technology allowing connectivity and trade in digital services becomes part of foreign policy and can be weaponised.

Therefore, conflicts over digital technology between the US and China should serve as an incentive for cooperation between the EU and South Korea to build a digital economy and an open and secure cyberspace with trustworthy technology, to increase their self-reliance and autonomy, and to support digital economies in the Indo-Pacific. This is in line with the EU Indo-Pacific strategy, which recognises explicitly the importance of a global, open, free and secure cyberspace.

The EU and South Korea have already developed distinctive cooperation in fields such as Research and Innovation (R & I), 5G, cloud computing and Internet of Things (IoT). The Joint Declaration on Strategic Cooperation in ICT and 5G, signed in 2014 by the European Commission Vice President responsible for the EU Digital Agenda and the South Korean Minister of Science, ICT and future Planning, was not only instrumental to developing joint R & I cooperation in the areas of fifth-generation mobile technology (5G), Cloud and IoT, but also paved the way for structured exchanges on 5G policy matters. The objective has been supporting global agreements on issues such as 5G standards.

Also in 2014, a Memorandum of Understanding was signed between the 5G Infrastructure Association (as part of the EU's 5G collaborative research programme) and South Korea's 5G Forum, providing the industrial support necessary to implement these policy commitments. The 5th EU-Korea Joint Science & Technology Cooperation Committee confirmed in 2015 5G and IoT as thematic area of cooperation of mutual interest. Already earlier in 2006, the EU and South Korea had signed an agreement for the joint development of Galileo, which is the EU-led global navigation-satellite-system designed as a backup for and strategic alternative to the American GPS. South Korea was the only East Asian country (in addition to China) to sign on. This may have set the scene for future information and communications technology (ICT) cooperation in space where the cybersecurity of satellite communication is a growing concern. The 7th Joint S & T Cooperation Committee Meeting in February 2022 covered principles for international cooperation, with Ethics, Gender and Open Science as examples and important work towards a digital partnership. South Korea's intention to start discussing its membership in the Horizon Europe programme as an associated country will facilitate cyber-related R & D.

The EU and South Korea have many compelling reasons to continue and intensify their cooperation in the field of ICT. First, the EU and South Korea are leaders in ICT: under the label '4th Industrial Revolution', both invest heavily in emerging tech such as Artificial Intelligence (AI), cloud computing, IoT and 5G. They are home to leading high-tech companies such as Ericson, Nokia, LG and Samsung. South Korea invests more in R & D than most countries in terms of GDP percentage (4.6% in 2019[1] according to the OECD). Furthermore, they share the same commitment to market economy, as evidenced by the free trade agreement (FTA) between the EU and South Korea which was in 2011 the EU's first free trade agreement in Asia. Lastly, the EU and South Korea are partners in building cyber capacity worldwide, including Internet connectivity for research and education communities across Asia-Pacific.

Second, on **cyber threats**: digital technology poses new and growing cyber threats which have to be met. While digital technology has revolutionised economies and societies in opening new opportunities, it created new security challenges such as cybercrime, violations of privacy, economic espionage and election interference. The scope of 'critical infrastructure' is

[1] https://data.oecd.org/rd/gross-domestic-spending-on-r-d.htm

growing because of digitalisation and constantly increases the realm of security policy: weaponisation and securitisation go hand-in-hand.

The EU and South Korea are frequent targets of sophisticated cyber-attacks. Russia is often cited as the EU's biggest cyber threat. In addition, the frontline of the US-China tech conflict goes through Europe as a popular target for espionage of both. South Korea is targeted particularly by Russia and North Korea. The hacking of government computers at the opening of the 2018 Winter Olympic Games in South Korea was a particularly noticeable and eye-opening attack.

To respond to cyber threats, many States have heavily invested in military and law enforcement capabilities, which help to fight crime and protect national security. At the same time there is the dark side of cyber, opening the door for human rights violations, hate speech and misinformation. A cyber arms race is in the making. In order to remedy this situation, rules-based international cooperation is the third area of cooperation between the EU and South Korea.

Third, on **rules-based international cooperation**: investment in digital technology is a fertile ground for competition but also for cooperation to face common challenges. In military terms, competition in cyber technology translates into a cyber arms race. This danger can be mitigated by developing rules for international cooperation and introducing concepts of arms control into the cyber sphere.

Therefore, the EU and South Korea entertain regular cyber and ICT dialogues which demonstrate their shared belief in multilateralism, respect for fundamental norms and values such as democracy, rule of law, human rights and international law to which cultural differences are no obstacle because of their universal nature.

The EU and South Korea have developed strong regulations for data protection, culminating in December 2021 in the adequacy decision by the European Commission under the General Data Protection Regulation (GDPR). This allows transfer of personal data from the EU to South Korea guaranteeing a comparable level of protection of personal data to that in the EU.

The Council of Europe's cybercrime Convention (Budapest Convention), a benchmark convention and source of inspiration for cyber-crime legislation worldwide, is an example of advanced rules-based cooperation. South Korea's accession would therefore be a strong statement in support of multilateralism. However, as demonstrated in the difficult discussions at the United Nations (UN), there is a still a long way to go to advance international law for cyberspace especially in fields such as cyber arms control for cyber weapons.

In identifying the similarity and to some extent the complementarity of policies followed by the EU and South Korea, the authors of this volume deal with the most salient issues of cybersecurity and make a strong case for enhancing cooperation in deeds, not only in words, bilaterally and globally.

2 Aim of the Book

The primary objective of this volume is to tap the potential of EU-South Korea cooperation to build a peacetime regime for cyberspace based on shared norms, rules and values. Public awareness of EU-South Korea ICT cooperation is relatively low and has been neither fully appreciated nor used. Therefore, the editors launched at a first step workshops in South Korea and later due to the Covid-19 pandemic online with the goal to unite expertise from different fields for a publication. This led to a consortium believing in cyber peace and the need to pursue investments in cybersecurity in parallel with investments in democracy and a rules-based international order. The war in the Ukraine tragically underlines the relevance of this topic. The tome informs about the extent and intensity of ICT connections between the EU and South Korea, and explains in plain language state-of-the-art policy-oriented research about cybersecurity, thereby raising awareness for the need to invest in cyber peace.

Another objective of this book is to bring together, for the first time, scattered and fragmented knowledge and ongoing interdisciplinary research about cybersecurity in a structured and coherent manner. It is built on the conviction that experts from various continents and all areas involved have to join hands to achieve cybersecurity and cyber peace. Silo thinking no longer produces results. Therefore, this volume contains chapters on cybersecurity in terms of law, political science and international relations including security studies.

3 Target Audience

The book's main target professional audience consists of public sector policymakers, military and law enforcement officers, private sector cybersecurity experts, as well as lawyers, media and civil society specialised in cybersecurity. The authors anchored in academia also aim to attract the interest of university students and scholars in a variety of disciplines such as computer science, cognitive science, political science, international relations, international law, and strategic and defence studies.

This volume serves a wider audience as there is an increasing interest in a comprehensive understanding and policymaking in the field. This is also reflected by the introduction of teaching modules on cybersecurity at universities but also as part of vocational training across Europe and in South Korea, targeting students and government officials worldwide. Given its global outlook, the book hopes to draw the interest of technologically advanced countries as well as developing countries upgrading their economies with the help of smart and connected technologies, an area where South Korea has special knowledge and experience.

4 Focus and Outline

The focus is mainly on the EU and South Korea, though the work also takes into consideration the perspective of China, an important IT actor and neighbour of South Korea, and the Asian protagonist in the great power competition with the US, with direct repercussions for the European Union and work at the UN. The eight authors from the EU and the US meet seven Korean experts and one Chinese expert, complementing each other and laying proof that cyberspace and its security is a global endeavour. Where possible, the editors encouraged dialogues between contributors representing Korean and European perspectives.[2]

[2] All chapters highlight in bold the surnames of the authors to whom cross-references are made.

The division in three main sections shall facilitate navigation:

- *Section 1 'Rationale for cybersecurity cooperation between the EU and South Korea'* tackles the question 'why' the EU and South Korea should cooperate in the field of cybersecurity. Authors use clearly supporting arguments and often also robust comparisons, that is, a strong analysis of similarities and differences between the EU and South Korea. They identify lacunas to be filled through joint actions/ proposals. The themes covered by Section 1 are cyber diplomacy, data protection, cybersecurity, emerging technologies, offence-defence theory and cyber sanctions. Mason **Richey**'s contribution also includes a 'Cyber Threat Overview' with common cases of malicious cyber operations, which serves as a reference point throughout the book for other authors. The editors would like to specially thank Mason Richey for this extra endeavour.
- *Section 2 'Comparisons'* covers in six papers three themes (cyber arms control, cyber defence and digital investigations). Reflecting the endeavours of the editors, each of the three themes is dealt with by two authors offering a Korean and European perspective respectively.
- *Section 3 'International order and cross-cutting issues'* moves the analysis beyond the EU-South Korean angle identifying cross-cutting issues in the field of cybersecurity and the ongoing efforts with the UN system to advance international cybersecurity governance. China's position is presented to show the differences in approach in this process.

5 Outline of the Chapters

For the readers' convenience, the following table offers a short overview of the themes covered in each section by the respective author.

1. ***Introduction***

<u>Authors:</u> Gertjan BOULET, Michael REITERER, Ramon Pacheco PARDO

The introduction says why are you reading what you are reading, reflecting the shared commitment of the editors and authors to invest in three areas for cyber cooperation between the EU and South Korea: digital economy, cyber treats and cooperation to maintain or create a rules-based cyber world. The introduction also explains the book's aim, target audience, focus and outline.

Section 1. Rationale for Cybersecurity Cooperation Between the EU and South Korea

Theme	Chapter
Cyber diplomacy	2. ***EU Cyber Diplomacy: Part of a Value- and Interest-Driven Foreign Policy with New Focus on the Indo-Pacific*** <u>Author:</u> Michael REITERER This chapter focuses on the scope and three important functions of the EU cyber diplomacy: supporting externally the relevant internal EU policies and providing protective as well as offensive tools through the EU's Cyber Diplomacy Toolbox. The author also argues that the EU cyber diplomacy should act globally as one in the form of a comprehensive and joint EU cyber diplomacy reflecting European interests and values.
Data protection, cybersecurity, emerging technologies	3. ***EU-South Korea Cooperation on Cybersecurity, Data Protection and Emerging Technologies*** <u>Authors:</u> George CHRISTOU & Ji Soo LEE This chapter interrogates how cooperation and collaboration has evolved between the EU and South Korea in the areas of cybersecurity, data protection and emerging technologies. It is argued that whilst there is a high degree of convergence across the issue areas explored, divergence still exists and cooperation can be significantly improved beyond dialogue, information sharing and trust-building.
Offence-defence theory (middle powers amidst strategic competition)	4. ***Cyber Offence Dominance, Regional Dynamics, and Middle Power–Led International Cooperation*** <u>Author:</u> Mason RICHEY This chapter provides a cyber threat landscape and analyses strategic competition and conflict among selected states in East Asia (the US, China, South Korea, North Korea, Japan, Russia) via the cyber domain. Following that, this chapter applies offence-defence theory to the cyber domain in order to predict states' likely cyberspace behaviour. To this end, the chapter argues that the cyber domain is offence-dominant, leading to increased risks of escalatory chain-ganging. Second, flowing from this analysis, the chapter seeks to determine what kinds of international cooperation could reduce this escalatory cyber tendency, and how the EU is cooperating with like-minded states such as South Korea to reduce cyber threats.

Cyber sanctions	5. ***Developing a Collective Retorsion Framework Against Malicious Cyber Operations: Opportunities and Steps for EU-South Korea Cybersecurity Cooperation*** Author: Joohee PARK This chapter develops the argument that South Korea should seek cooperation with the EU to build a framework for collectively imposing proportionate costs on malicious cyber actors, a collective retorsion framework which is comprised of lawful but unfriendly acts. The author argues that the EU's restrictive measures promoted through the Framework for a Joint EU Diplomatic Response to Malicious Cyber Activities can offer useful insights into South Korea's cybersecurity policies.	
Section 2. Comparisons		
Theme	**South Korea**	**EU**
Cyber arms control	6. ***Cyber Arms Control on the Korean Peninsula: Challenges and Opportunities with a View to North Korea's Asymmetry Strategy*** Authors: Sangho LEE & Maximilian ERNST This chapter focuses on arms control in South Korea and investigates the possibility of a cyber arms control regime on the Korean Peninsula. Fashioned after existing arms control regimes, this chapter proposes an inter-Korean cyber arms control agreement.	7. ***Arms Control in the Cyber Domain: A European Approach to Mitigate Digital Threats*** Author: Michael ZINKANELL This chapter focuses on arms control in the EU and analyses the EU cybersecurity strategies and mechanisms from an arms control perspective. The chapter concludes that the EU has effectively developed and implemented instruments to successfully mitigate digital threats, indicating aspects of arms control and non-proliferation. The chapter also stresses that the EU is immediately responsible for preventing the multifaceted risks and threats posed by artificial intelligence.

(*continued*)

(continued)

Cyber defence	8. ***South Korea's Cyber Defence and Digital Cooperation with the EU*** Authors: Kyu-dok HONG & Seong-jong SONG This chapter focuses on cyber defence and resilience in South Korea. The chapter examines the status of cyber threats stemming from North Korea and explores ways to strengthen cyber defence between South Korea and the EU. The authors formulate policy recommendations to improve South Korea's cybersecurity defence posture and its relevance for further cyber cooperation between South Korea and the EU.	9. ***Building Cyber Resilience: The Defensive Shield for the EU*** Author: Anna-Maria OSULA This chapter focuses on cyber defence and resilience in the EU. This chapter explains in a nutshell the EU's main regulatory and policy aims in achieving increased resilience in cyberspace. In particular, the chapter focuses on the role of disruptive technologies in building a defensive shield for the EU and recent developments within the domain of EU cyber defence. The author explains how cyber resilience builds on the EU's existing regulatory framework, yet that Brussels should also focus on developing pan-EU cyber defence capabilities. Finally, the chapter illustrates how cooperation with trusted partners such as South Korea will be mutually beneficial for enhancing cyber resilience.

Digital investigations	10. ***Enhancing Cooperation Between South Korea and the EU in the Fight Against Cybercrime*** Author: Gibum KIM This chapter reviews South Korea's international cooperation framework on cybercrime and presents four solutions to promote cooperation between South Korea and the EU: (1) joining the Convention on Cybercrime of the Council of Europe, (2) establishing a working group on joint investigation in cybercrime, (3) implementing cybercrime capacity building projects in developing countries, (4) expanding cooperation on digital forensics technology. The author argues that through these solutions, South Korea will be able to respond to the global cybercrime threat more actively and take more responsibility for global cybersecurity.	11. ***The EU and Access to Electronic Evidence: Privatisation of Law Enforcement?*** Author: Tatiana TROPINA This chapter discusses the challenges of current regimes governing access to electronic evidence stored with providers abroad and analyses the new legislative frameworks relevant to the EU and their possible effects. The author argues that replacing traditional mutual legal assistance mechanisms with direct requests to providers can lead to privatisation of law enforcement and, ultimately, have a severe effect on the protection of fundamental rights in the EU and beyond.
Section 3. International Order and Cross-cutting Issues		
UN cybersecurity governance	12. ***Recent Trends in UN Cybersecurity Governance and South Korea-EU Cooperation*** Author: JoonKoo YOO The chapter discusses why the UN Group of Governmental Experts (UNGGE) on Developments in the Field of Information and Telecommunications process has reached a turning point. It argues that the discussion about how Information and Communication Technology should be regulated is as much about strategy, politics and ideological differences as it is about law. This chapter also offers suggestions about what the future holds with regard to the regulation of cyberspace. It argues that the deadlock of the UNGGE process is likely to lead to a shift away from ambitious global initiatives towards regional cooperation between like-minded countries. In this regard, it delineates how to facilitate cooperation between the EU and South Korea in the context of governance and norms building.	

(*continued*)

(continued)

International Cybersecurity Cooperation	13. ***The International Cybersecurity Cooperation Dilemma and Implications for EU-South Korea Relations*** Author: Cuihong CAI This chapter focuses on the international cybersecurity cooperation dilemma. The author argues that international cybersecurity cooperation is still insufficient, and that EU-South Korea cybersecurity cooperation should be open and inclusive to eliminate misperceptions, develop a tailor-made framework for cybersecurity cooperation, establish a dedicated cybersecurity cooperation mechanism and pursue greater autonomy in the cyberspace of confrontation between camps. The author finds that there is room in their cooperation for the EU and South Korea to play a greater role in cybersecurity.
14. ***Conclusions: From Words to Deeds*** Authors: Michael REITERER, Gertjan BOULET, Ramon Pacheco PARDO In light of the speed of technological developments and the challenges posed for critical infrastructure, the concluding chapter sends a clear invitation to move cybersecurity from words to deeds. Based on an interdisciplinary and intercultural analysis of the chapters in the volume, the conclusions identify areas of cooperation between the EU and South Korea, anchored in common values and therefore aiming at an open, secure, rules-based cyber world.	

Acknowledgements The editors would like to thank, first, the authors for their smooth cooperation and contribution at an unpredictable time overshadowed by the covid-19 pandemic.

Special recognition for Mason Richey who volunteered to include a 'Cyber Threat Overview' in his chapter, which serves as an invaluable reference point throughout the book.

The editors would also like to express gratitude to all the professors and universities for supporting the inception of this volume by hosting one or more workshops or guest lectures: Professor Hyoung Joong Kim (Korea University, School of Cybersecurity), Professor Hun-Yeong Kwon (Korea University, School of Cybersecurity), Professor Yong Hak Kim (Yonsei University, former President), Professor Zoonky Lee (Yonsei University, Dean of the Graduate School of Information Security) and Professor Beomsoo Kim (Yonsei University, Graduate School of Information Security).

On 11 June 2018, Korea University had invited Ambassador Dr Michael Reiterer (then Ambassador of the European Union to the Republic of Korea) as a guest speaker at the graduate course on 'advanced cyber law' taught by Gertjan Boulet. Michael Reiterer presented on the 'EU's rules and law-based foreign policy for cyberspace'.

This discussion brought to the fore the similarities in approach by the EU and South Korea in building a cyber regime based on shared law, norms and values. These led to further workshops to analyse various aspects of international cybersecurity.

- The first workshop (4 July 2018) hosted by Michael Reiterer, Ramon Pacheco Pardo and Gertjan Boulet at the residence of Michael Reiterer set the scene for this volume.
- The second workshop (25 March 2019) was hosted by Professor Beomsoo Kim (Yonsei University), in the context of the State Visit of Their Majesties the King and the Queen of Belgium to South Korea (25–28 March 2019). A high-level Vrije Universiteit Brussel (VUB)-delegation joined the seminar, with congratulatory remarks by the Vice-rector International Relations, and the active participation of the Executive Director of the Brussels School of Governance (BSoG-VUB) and the head of the Department of Communication Studies.

- The third workshop (17 September 2019) was again hosted by Professor Beomsoo Kim at Yonsei University.
- The fourth workshop (28 July 2020), hosted by Michael Reiterer at his Residence, served to take stock with Korean authors and offered the occasion to bid farewell for the Ambassador upon completion of his assignment in August 2020.
- The fifth (final) workshop (7 September 2021) was held online by the editors at the Brussels School of Governance (BE), giving authors the chance for mutual feedback in the final phase of drafting chapters.

Last but not least, the editors would like to thank the anonymous reviewers of the book proposal, Palgrave's editors (Sarah Roughley, Rebecca Roberts, Alina Yurova, Arunaa Devi, Ulrike Stricker, Sarulatha Krishnamurthy, and Sylvia Anand) and the editor of the New Security Challenges Series (George Christou) for their constructive and invaluable feedback and their trust in the authors to provide innovative insights to this volume.

PART I

Rationale for Cybersecurity Cooperation Between the EU and South Korea

EU Cyber Diplomacy: Value- and Interest-Driven Foreign Policy with New Focus on the Indo-Pacific

Michael Reiterer

1 Introduction

Foreign policy reflects domestic politics. As digital technology becomes omnipresent in political, economic and social life, the security implications of its use increase. Cyber has become an integral part of life and infrastructure to the extent that it is necessary for their functioning—the term "critical infrastructure" ("critical" for the functioning of a society/economy) expresses this importance well. Cybersecurity is not only an integral aspect of security policy, whether domestic or international, but also of international relations in general.

M. Reiterer (✉)
Centre for Security, Diplomacy & Strategy (CSDS), Brussels School of Governance (BSoG-VUB), Brussels, Belgium
e-mail: michael.reiterer@vub.be

© The Author(s), under exclusive license to Springer Nature Switzerland AG 2022

G. Boulet et al. (eds.), *Cybersecurity Policy in the EU and South Korea from Consultation to Action*, New Security Challenges,
https://doi.org/10.1007/978-3-031-08384-6_2

Therefore, cyber diplomacy[1] is no longer an add-on, but a form of diplomacy in its own right. As a policy area it is focused on economic, security and political aspects of cyberspace and is therefore different from digital diplomacy which considers information and communications technologies (ICTs) as one of the tools of diplomacy.[2] Digital diplomacy requires a secure digital environment, that is, a safe and secure cyberspace which is the realm of cyber diplomacy.

The importance of cyber is reflected in concerns about cyberwarfare, cyber defence, cyber sabotage, cyber theft, cyber espionage, and cyber-based weapon systems. The increased ability and willingness of state and non-state actors to pursue their objectives through malicious cyber activities pose a threat to modern societies and may constitute wrongful acts under international law.

Already in the past, digital technology has become an enabler for non-state actors: positive, if used for participatory democracy, or spreading information on how to deal with the covid-19 pandemic; negative, in terms of fermenting support for extremism, hate and violence. The ever-increasing technological capabilities of the emerging high tech add an additional and defining aspect to "critical infrastructure" and the need for its protection as cyber attacks can impact on the whole economy, infrastructure, and societies.

The Fourth Industrial Revolution with Big Data and cloud-based technology will give connectivity a new dimension. In times of geopolitics and geo-economics, the EU is aware of the need to integrate cybersecurity into its foreign policy and to engage with third country partners. Thus, EU Foreign Ministers started to discuss the geopolitics of new digital technologies in July 2021.[3] Control of high speed, low latency and great capacity of the fifth generation (5G) of wireless technology has become essential. Developers may compromise its security through inbuilt entrance doors in particular when companies like Huawei are under the obligation to report on their activities. This drew the attention of the larger public to the danger of the use of compromised components in the production

[1] Shaun Riordan and Mario Torres Jarrín, *Global Policy Perspective Report: Cyberdiplomacy* (Salamanca-Stockholm: European Institute of International Studies, 2020), 4, http://www.ieeiweb.eu/wp-content/uploads/2020/01/Cyberdiplomacy_enero.pdf .

[2] "Digital diplomacy | E-diplomacy | Cyber diplomacy," DiploFoundation, accessed February 2, 2020, https://www.diplomacy.edu/e-diplomacy.

[3] "Foreign Affairs Council," Newsroom, EU Council, July 12, 2021, https://newsroom.consilium.europa.eu/events/20210712-foreign-affairs-council-july-2021.

process and for supply chain security. Thus, "a global race for leadership in key digital technologies or enabling systems ... is increasingly characterised by international tension and a growing 'geopoliticisation' of digital technologies around the globe."[4]

This chapter sets out the development of the EU's diplomatic approach to cybersecurity including its Cyber Diplomacy Tool (Sect. 2), explains the security-technology nexus (Sect. 3), highlights the importance of Asia and the Indo-Pacific as technology hubs and potential security risks (Sect. 4), puts the EU's endeavours in the context of the Liberal International Order (Sect. 5) and makes suggestions to upgrade the EU cyber diplomacy in the concluding Sect. 6.

2 Cybersecurity and Foreign Policy: Cyber Diplomacy

The EU 2013 Cybersecurity Strategy, 'An Open, Safe and Secure Cyberspace'[5] outlining how best to prevent and respond to cyber disruptions and attacks, marks the beginning of specific cyber activities. The Strategy also advocates the development of a cyber defence policy and capabilities as part of the Common Security and Defence Policy (CSDP).

2.1 *Cyber Diplomacy Toolbox*

Following the 2013 EU Cybersecurity Strategy the Council adopted in February 2015 Conclusions on Cyber Diplomacy[6] which still are guiding principles:

1. Promotion and protection of human rights in cyberspace, norms of behaviour and application of existing international law in the field of international security.

[4] European Political Strategy Centre (European Commission), ***Rethinking Strategic Autonomy in the Digital Age*** (EPSC Strategic Notes no. 30, July 2019), 2, https://op.europa.eu/en/publication-detail/-/publication/889dd7b7-0cde-11ea-8c1f-01aa75ed71a1/language-en/format-PDF/source-118064052.

[5] European Commission (EC) – Joint Communication, *Cybersecurity Strategy of the European Union: An Open, Safe and Secure Cyberspace* (Brussels: February 2, 2013), https://eeas.europa.eu/archives/docs/policies/eu-cyber-security/cybsec_comm_en.pdf.

[6] EU Council, *Council Conclusions on Cyber Diplomacy* (Brussels: February 15, 2015), http://data.consilium.europa.eu/doc/document/ST-6122-2015-INIT/en/pdf.

2. Internet governance.
3. Enhancing competitiveness and prosperity.
4. Capacity building and development.

The concretisation of these principles led the EU Foreign Ministers in 2017 to establish the "Cyber Diplomacy Toolbox" (CDT)[7] to strengthen the EU's joint response to malicious cyber activities that harm its political, security and economic interests.

This CDT has a double function in complementing domestic measures: (1) contributing to conflict prevention, mitigating cybersecurity threats and consequently greater stability in international relations; (2) deterring potential aggressors from attacking the EU through the threat of a common EU response including the imposition of restrictive measures, sanctions against individuals or entities.

A joint EU response would be proportionate to the scope, scale, duration, intensity, complexity, sophistication and impact of the cyber activity. Furthermore, states should not knowingly allow their territory to be used for internationally wrongful acts; any failure to prevent such actions could lead to a joint EU response.

In order to be credible, the EU established in 2019 a framework allowing "to impose **targeted** restrictive measures to deter and respond to cyber-attacks which constitute an external threat to the EU or its member states, including cyber-attacks against third States or international organisations where restricted measures are considered necessary to achieve the objectives of the Common Foreign and Security Policy (CFSP)" (*emphasis added*).[8]

Cyber attacks falling within the scope of this new sanction regime are those which have a *significant impact* and which

- Originate or are carried out from outside the EU.
- Use infrastructure outside the EU.
- Are carried out by persons or entities established or operating outside the EU.

[7] EU Council, *Draft Council Conclusions on a Framework for a Joint EU Diplomatic Response to Malicious Cyber Activities ("Cyber Diplomacy Toolbox")* (Brussels: June 7, 2017), http://data.consilium.europa.eu/doc/document/ST-9916-2017-INIT/en/pdf.

[8] "Cyber-attacks: Council is now able to impose sanctions," Press Releases, EU Council, May 17, 2019, https://www.consilium.europa.eu/en/press/press-releases/2019/05/17/cyber-attacks-council-is-now-able-to-impose-sanctions/.

- Are carried out with the support of person or entities operating outside the EU.

Attempted cyber attacks with a potentially significant effect are also covered by this sanction regime. Only one year later, in July 2020, the EU used this tool for the first time and imposed a travel ban and an asset freeze on eight individuals and four entities.[9]

In order to increase the deterrence value, the CDT states that an attack on the EU would result in a common response:

> "4. The EU stresses that clearly signalling the likely consequences of a joint EU diplomatic response to such malicious cyber activities influences the behavior of potential aggressors in cyberspace thus reinforcing the security of the EU and its Member States. The EU reminds that attribution to a State or a non-State actor remains a sovereign political decision based on all-source intelligence and should be established in accordance with international law of State responsibility. In that regard, the EU stresses that not all measures of a joint EU diplomatic response to malicious cyber activities require attribution to a State or a non-State actor."[10]

Therefore, cybersecurity is also a longstanding topic of EU-NATO cooperation as they are faced with similar threats undermining "all levels of society in member states, threatening civil, political, economic and military security. The vast increase in the number of cyberattacks and the emergence of cyberspace as a new battlefield has motivated the EU and NATO to launch initiatives to strengthen their cyber resilience, and to increase their mutual consultations and coordination in the cyber domain."[11]

[9] "EU imposes the first ever sanctions against cyber-attacks," Press release, EU Council, July 30, 2020, https://www.consilium.europa.eu/en/press/press-releases/2020/07/30/eu-imposes-the-first-ever-sanctions-against-cyber-attacks/.

[10] EU Council, *Draft*, 5, para. 4.

[11] Bruno Lété, "Cooperation in cyberspace," in *The EU and NATO: the essential partners,* ed. Gustav Lindstrom and Thierry Tardy (Paris: European Union Institute for Security Studies, 2019), 28; "European Union and NATO work together to tackle growing cyber threats," European Union External Action Service (EEAS), December 11, 2018, https://eeas.europa.eu/topics/economic-relations-connectivity-innovation/55217/european-union-and-nato-work-together-tackle-growing-cyber-threats_en.

NATO Secretary General Stoltenberg proclaimed that a serious cyber attack on one member of NATO could lead to the activation of Art. 5 of the Alliance, the "collective defence commitment":

> For NATO, a serious cyberattack could trigger Article 5 of our founding treaty. This is our collective defence commitment where an attack against one ally is treated as an attack against all. We have designated cyberspace a domain in which NATO will operate and defend itself as effectively as it does in the air, on land, and at sea. This means we will deter and defend against any aggression towards allies, whether it takes place in the physical world or the virtual one.[12]

Cybersecurity and related threats are in the category of hybrid threats.[13] To enhance legal resilience in implementing joint activities between the EU and NATO, the EU had set up in 2017 the European Centre of Excellence for Countering Hybrid Threats in Helsinki.[14] Mapping cyber threats is part of the tasks of the Centre.

In order to foster security in Europe the 2020 EU Security Union Strategy lays out the tools and measures to be developped until 2025. They range "from combatting terrorism and organised crime, to preventing and detecting hybrid threats and increasing the resilience of our critical infrastructure, to promoting cybersecurity and fostering research and innovation".[15] To this end the EC proposed "a new Joint Cyber Unit to tackle the rising number of serious cyber incidents impacting public services, as well as the life of businesses and citizens across the European Union".[16] This was confirmed by the Council in October 2021 inviting

[12] BBC News, "NATO: Cyber-attack on one nation is attack on all," *BBC News*, August 27, 2019, https://www.bbc.com/news/technology-49488614.

[13] Daniel Fiott, *The Yearbook of European Security 2019* (Paris: EUISS, 2019), 92-93 and Annex, https://www.iss.europa.eu/sites/default/files/EUISSFiles/yes2019.pdf.

[14] "Home", The European Centre of Excellence for Countering Hybrid Threats (HybridCoE), accessed July 21, 2021, https://www.hybridcoe.fi/.

[15] EC, *Communication from the Commission to the European Parliament, the European Council, the Council, the European economic and social Committee and the Committee of the regions – on the EU Security Union Strategy* (Brussels: July 24, 2020), https://eur-lex.europa.eu/legal-content/EN/TXT/PDF/?uri=CELEX:52020DC0605&from=EN.

[16] EC, "EU Cybersecurity: Commission proposes a Joint Cyber Unit to step up response to large-scale security incidents," Press release, June 23, 2021, https://ec.europa.eu/commission/presscorner/detail/en/IP_21_3088.

member states to identify information sharing gaps and needs within and across cyber communities.[17]

The growing weaponisation of trade policy by major players is a wake-up call in this respect, as it leads to the disruption of not only e-commerce but also supply lines causing disruptions in technology, research and materials (semi-conductors) needed in the cyber-related production processes. A large trade deficit for high-end products,[18] especially with China, and including critical raw materials, indicates the need for corrective policy actions for securisation.

Working on security, the Permanent Structured Cooperation (PESCO) includes "projects designed to enhance the EU's critical defence capabilities",[19] including the *Cyber Threats and Incident Response Information Sharing Platform* and the *Cyber Rapid Response Teams* and *Mutual Assistance in Cyber Security*.[20] In 2018 the EU Cyber Defence Policy Framework (CDPF)[21] added training, research and technology, civil-military cooperation and international cooperation to the 2014 scheme which had aimed at developing cyber defence capabilities including CSDP communication and information networks.

The 2020 New Cyber Security Strategy[22] proposal is part of a larger package which includes the above-mentioned EU Security Union Strategy (2020–2025),[23] the Digital Single Market and the Economic Recovery

[17] Council Press Release (Brussels 19 October 2021). *Cybersecurity: Council adopts conclusions on exploring the potential of a joint cyber unit*. https://www.consilium.europa.eu/en/press/press-releases/2021/10/19/cybersecurity-council-adopts-conclusions-on-exploring-the-potential-of-a-joint-cyber-unit/.

[18] EPSC, *Rethinking*, 8–9.

[19] Daniel Fiott, "Strategic autonomy: towards 'European sovereignty' in defence?," *EUISS Brief* 12 (2018), 6, https://www.iss.europa.eu/content/strategic-autonomy-towards-%E2%80%98european-sovereignty%E2%80%99-defence.

[20] EU Council, *Decision establishing the list of projects to be developed under PESCO* (Brussels: March 1, 2018), http://data.consilium.europa.eu/doc/document/ST-6393-2018-INIT/en/pdf.

[21] EU Council, *EU Cyber Defence Policy Framework (2018 update)* (Brussels: November 19, 2018), https://data.consilium.europa.eu/doc/document/ST-14413-2018-INIT/en/pdf.

[22] EC – Joint Communication to the European Parliament and the Council, *The EU's Cybersecurity Strategy for the Digital Decade* (Brussels: December 16, 2020), https://digital-strategy.ec.europa.eu/en/library/eus-cybersecurity-strategy-digital-decade-0.

[23] EC, *EU Security Union Strategy* (July 24, 2020).

Plan[24] to promote digitalisation, and to increase the quality and resilience of the EU cyberspace.

This New Strategy has a strong focus on international norm and standard setting through global cooperation for a global, open, stable and secure cyberspace while strengthening the cyber and physical resilience of critical infrastructure. With the proposed Directive on the resilience of critical entities[25] in ten sectors[26] an all-hazards framework is set up to enable critical entities to prevent, resist, absorb and recover from any disruptive incidents.

In March 2021 the Council,[27] recalling that the common and comprehensive cyber diplomacy aims at "contributing to conflict prevention, mitigation of cybersecurity threats and greater stability in international relations", adapted cybersecurity to the new challenges posed by covid-19 and the ensuing new policy to achieve a resilient, green and digital Europe[28] with the goal of strategic autonomy. The 2022 Work Programme of the European Commission[29] followed up in planning a Cyber Resilience Act to introduce common cybersecurity standards for connected devices.

Concerning strategic autonomy, the close link between military and civil applications necessitates further strengthening of the EU's cyber resilience:[30] The command of the relevant emerging critical technologies promoted by an industrial policy as part of security policy will provide a

[24] "Recovery Plan for Europe," European Commission, accessed August 10, 2021, https://ec.europa.eu/info/strategy/recovery-plan-europe_en.

[25] EC, *Proposal for a Directive of the European Parliament and of the Council on the resilience of critical entities* (Brussels, 16 December 2020), https://ec.europa.eu/home-affairs/system/files/2020-12/15122020_proposal_directive_resilience_critical_entities_com-2020-829_en.pdf.

[26] These sectors are energy, transport, banking, financial market infrastructures, health, drinking water, waste water, digital infrastructure, public administration, space.

[27] EU Council, *Draft Council conclusions on the EU's Cybersecurity Strategy for the Digital Decade* (Brussels, 9 March 2021), https://data.consilium.europa.eu/doc/document/ST-6722-2021-INIT/en/pdf.

[28] The Digital Europe Program devotes 269 million out of 1.98 billion euros to improving cybersecurity. EURAXESS, Press Release 12 November 2021, https://euraxess.ec.europa.eu/worldwide/south-korea/commission-invest-nearly-%E2%82%AC2-billion-digital-europe-programme-advance.

[29] EC. *Commission work programme 2022 – Making Europe stronger together*, (Brussels, 19 October 2021), https://ec.europa.eu/info/sites/default/files/com2021_645_en.pdf.

[30] EPRS, *Briefing*.

solid industrial base. This will serve as a pillar of cyber resilience, thereby reducing vulnerability.

2.2 *Fighting Disinformation*

Combatting disinformation has become a necessary task to safeguard trust in and resilience of democracy. To this end the EU has concluded a Code of Practice against disinformation with the three online platforms Facebook, Google and Twitter[31] and set up a special task force as part of the Strategic Communications and Information Analysis Division of the European External Action Service (EEAS).[32] This is also crucial in the systemic competition between authoritarianism and open democratic societies.

3 The Security-Technology Nexus

Technological competition is particularly salient in the fast-developing cyber sector. The use or ban of 5G technology originating in non-like-minded countries like China is perceived by many as a security threat which needs to be addressed by cyber diplomacy.

In order to remain a rule-maker and not become a rule-taker the EU needs to get ahead of the technological curve of cyber technology. A lack of knowledge and loss of leadership in this area would present a danger to the security of critical infrastructure and essential services.[33] This adds an important dimension to cyber diplomacy in content and scope.

Technical preparedness for cyber attacks is one side of the coin. The other side is the capacity to defend and respond in an adequate and proportional manner, applying due diligence primarily in a political manner as

[31] "Code of Practice against disinformation: Commission recognises platforms' efforts ahead of the European elections," Press Corner, European Commission, May 17, 2019, https://europa.eu/rapid/press-release_STATEMENT-19-2570_en.htm; See also Mélanie Scheidt, "The European Union versus External Disinformation Campaigns in the Midst of Information Warfare: Ready for the Battle?," *EU Diplomacy Papers* no 1, (August 2019), https://www.coleurope.eu/study/eu-international-relations-and-diplomacy-studies/research-publications/eu-diplomacy-papers.

[32] EEAS. *Questions and Answers about the East StratCom Task Force* News stories (Brussels 28 April 2021), https://eeas.europa.eu/headquarters/headquarters-homepage/2116/-questions-and-answers-about-the-east-stratcom-task-force_en.

[33] EPSC, *Rethinking*, 3.

laid down in the CDT. Counter hacks such as attempts to destroy a server which has been established as the source for the cyber attack could amount to an escalation of the response.[34]

5G technology plays a particularly important role as it may develop into the "de facto 'central nervous system' of the economy".[35] 5G and its successor networks will be a major enabler for future digital services in core areas of citizens' lives and an important basis for digital and green transformations. With worldwide annual 5G revenues estimated at €225 billion in 2025, 5G is a key asset for Europe to compete in the global market; its cybersecurity is crucial for ensuring the strategic autonomy of the Union.

Therefore, the EC highlighted in its 2019 Recommendation on Cybersecurity of 5G networks[36] as well as in the Strategic Outlook on China[37] the need to achieve a high level of security. Software-based applications, such as operating systems and cloud computing, dominated presently by five US companies[38], will need policy attention to avoid over-dependencies or opening the door to blackmailing. When setting up the European satellite system, Galileo, to avoid total dependence on the US GPS, the EU acted strategically, establishing a precedent to follow.

Consequently, in addition to the CDT, a specific 5G toolbox[39] was adopted in 2020 to mitigate the many risks identified in the 2019 Risk Assessment Report. They include the risk of interference from non-EU state or state-backed actors in making use of the 5G supply chain. In terms of foreign policy, member states agreed to strengthen security requirements, to assess the risk profiles of suppliers, to apply relevant restrictions for suppliers considered to be high risk including necessary exclusions for

[34] Annegret Bendiek, "The EU as a Force for Peace in International Cyber Diplomacy," *SWP Comment* no. 19 (April 2018), https://www.swp-berlin.org/fileadmin/contents/products/comments/2018C19_bdk.pdf.

[35] EPSC, *Rethinking*, 4.

[36] European Commission, "Commission Recommendation of 26.3.2019 Cybersecurity of 5G networks," Library, March 26, 2019, https://digital-strategy.ec.europa.eu/en/library/cybersecurity-5g-networks.

[37] EC and HRVP contribution to the European Council, "EU-China – A strategic outlook," March 12, 2019, https://ec.europa.eu/info/sites/default/files/communication-eu-china-a-strategic-outlook.pdf.

[38] EPSC, *Rethinking*, 9.

[39] NIS Cooperation Group, *Cybersecurity of 5G networks EU Toolbox of risk mitigating measures* (CG Publication, January 29, 2020), https://ec.europa.eu/digital-single-market/en/news/cybersecurity-5g-networks-eu-toolbox-risk-mitigating-measures.

key assets considered as critical and sensitive (such as the core network functions), and to have strategies in place to ensure the diversification of vendors.

4 Asia and the Indo-Pacific in the Limelight

Cybersecurity at the intersection between security and technology poses continuously evolving challenges for the EU's cyber diplomacy. Particular parts of Asia which are home to important IT industry producers are concerned. Asia is also home to geographical/territorial tensions where interstate conflicts can easily heat up and spill over into cyberspace. Thus, Asia needs diplomatic attention as a technology hub as well as a potential security risk.

In recognition of these risks the EU Council adopted the 2018 Conclusions on Enhanced EU security cooperation in and with Asia.[40] These conclusions identify cybersecurity as one of the key areas for deeper security engagement, together with maritime security, counter terrorism, hybrid threats, conflict prevention, the proliferation of chemical biological, radiological and nuclear weapons, and the development of regional cooperative orders.[41]

South Korea is the pilot country for cyber cooperation.[42] At the virtual June 2020 EU-South Korea meeting leaders confirmed the need for enhancing cooperation for digitalisation which should be human-centred; they also committed to launch a High-Level Dialogue on the digital economy.[43] Such a dialogue can help each partner to strike the right balance between individual data protection/common health care, competitivity/security, to name a few difficult choices.

[40] EU Council, ***Enhanced EU Security Cooperation in and with Asia – Council conclusion (9265/1/18 REV 1)*** (Brussels: May 28, 2018), http://www.consilium.europa.eu/media/35456/st09265-re01-en18.pdf.

[41] EU Council, ***Enhanced EU Security***, Annex para. 6.

[42] GIZ, ***Enhancing Security Cooperation in and with Asia***, accessed 28 October 2021, https://www.giz.de/en/worldwide/87412.html. See also **Hong and Song** in this volume.

[43] EU Council, ***Joint press release: Republic of Korea – EU Leaders' video conference meeting***, 30 June 2020, accessed 7 October 2020, https://www.consilium.europa.eu/en/press/press-releases/2020/06/30/joint-press-release-republic-of-korea-eu-leaders-video-conference-meeting/.

The EU's 2018 Connectivity Strategy[44] adds comprehensive "connectivity partnerships", which have a strong cyber component, to the EU policy tool box. The Strategy strives to build networks that bring people, places and opportunities closer together, with a particular focus on transport, energy, digital and human-to-human connectivity. The Global Gateway[45] enlarges this tool in significance and geographical scope like the 2021 Indo-Pacific Strategy.[46]

The 2019 ASEAN-EU Statement on Cybersecurity Cooperation agreed at the ASEAN-EU Ministerial Meeting encapsulates the will to cooperate recognising in particular "that cyberspace provides opportunities for social and economic development, but also poses continuously evolving challenges to our regions that may harm international peace and security".[47] EU cyber diplomacy needs to work with like-minded partners to establish collective, open cybersecurity regimes. The competition is on: ASEAN entertains various other dialogues and platforms such as with the US, or ASEAN Defence Ministers Plus Working Group on Cybersecurity, or a Senior Cyber Group with Quad.

In April 2021 the Council made cybersecurity one of the priorities of the EU′s Strategy for Cooperation in the Indo-Pacific (IP) to promote digital governance and standard setting, to assure a global, open, free, stable and secure cyberspace for continued prosperity and protection of fundamental human rights in cyberspace.[48] Cybersecurity is singled out as the key topic for security and defence dialogues with partners. To this end

[44] EC – Joint Communication, *The EU Strategy Connecting Europe and Asia – Building blocks for an EU Strategy* (Brussels: September 19, 2018), https://eeas.europa.eu/sites/eeas/files/joint_communication_-_connecting_europe_and_asia_-_building_blocks_for_an_eu_strategy_2018-09-19.pdf.

[45] Joint Communication to the European Parliament, the Council, the European Economic and Social Committee, the Committee of the Regions and the European Investment Bank (2021). *The Global Gateway.* JOIN (2021) 30 final, 1 December 2021; at https://ec.europa.eu/info/sites/default/files/joint_communication_global_gateway.pdf.

[46] EC-Joint Communication to the European Council and European Parliament, *The EU strategy for cooperation in the Indo-Pacific* (Brussels, September 16, 2021) https://eeas.europa.eu/headquarters/headquarters-homepage/104126/joint-communication-indo-pacific_en.

[47] ASEAN, *ASEAN-EU Statement on Cybersecurity Cooperation* (Thailand: 2019), https://asean.org/storage/2019/08/ASEAN-EU-Statement-on-Cybersecurity-Cooperation-FINAL.pdf.

[48] EU Council, *EU Strategy for cooperation in the Indo-Pacific* (Brussels: April 16, 2021), 10, https://data.consilium.europa.eu/doc/document/ST-7914-2021-INIT/en/pdf.

a "EU Cyber Diplomacy Network, working with the EU Delegations, as well as with relevant member states' embassies around the world" was announced.[49]

Thereby, the EU's cyber diplomacy strives to build digital partnerships "to foster inclusive growth, tackle inequalities and transform the delivery of public services, enabling the achievement of the 2030 Agenda and the Sustainable Development Goals (SDGs) ... but (which) also bear the risk of deepening existing inequalities and being misused to undermine democracies and social cohesion and infringe on human rights".[50] This approach is built on "a human-centric vision for the digital economy and society ... a technology (that) works for the people", that creates digital economies which "are fair and inclusive", as part of open, democratic and sustainable societies.[51] In doing so, the EU can build on ongoing cooperation with partners in the Asia Pacific. In implementing the EU-Singapore Trade Agreement and attempting to bring it into the digital future, both parties strive to "strengthening bilateral digital trade, including with a view to advancing towards a comprehensive EU-Singapore digital partnership".[52] Singapore is already a member of the Digital Economy Partnership Agreement (DEPA) with Chile and New Zealand,[53] which is set to expand (South Korea is considering joining it too) and can serve as a source of inspiration. Since 2009 there is an ICT Dialogue to which in 2012 the EU-China Cyber Task Force was added. The Indo-Pacific dimension of cyber diplomacy, including Africa Connect and the Digital4Development Hub, was confirmed by the declaration issued by the "Forum for Cooperation in the Indopacific" organised by the French

[49] EC and HRVP, *The EU strategy for cooperation in the Indo-Pacific* (Brussels: September 21, 2021), 15, https://ec.europa.eu/info/sites/default/files/jointcommunication_indo_pacific_en.pdf.

[50] EC. Global digital partnerships; at https://ec.europa.eu/international-partnerships/topics/digital-partnerships_en (accessed 26 February 2022).

[51] EC-International Partnerships, Responsible digitalisation, accessed 26 February 2022, https://international-partnerships.ec.europa.eu/policies/digital-and-infrastructure/responsible-digitalisation_en#related-documents.

[52] EC (2021). Joint Statement: EU and Singapore agree to strengthen bilateral partnership on digital trade Press release, 7 December 2021; at https://trade.ec.europa.eu/doclib/press/index.cfm?id=2337.

[53] Ministry of Trade and Industry (MTI) of Singapore. Digital Economy Partnership (DEPA); at https://www.mti.gov.sg/Improving-Trade/Digital-Economy-Agreements/The-Digital-Economy-Partnership-Agreement (accessed 26 February 2022).

Presidency of the Council of the European Union in Paris, 22 February 2022.[54]

5 Cyber Issues and the Liberal International Order

The EU's promotion of "an open, free, stable and secure cyberspace" is part of its policy to support the Liberal International Order (LIO) and the principles and rules it stands for.

International cooperation is a necessity in the face of the weakening of multilateralism, growing nationalism and authoritarianism and accelerating weaponisation of policies including in the digital and cyber spheres, and space (critical role of satellites), which offer unprecedented dangers to critical infrastructure.

The discussion about states' sovereignty in cyberspace is intensifying while progress towards a stable regime for the digital domain—based on norms, international law, and confidence building measures—remains slow. This is hardly surprising as many states regard cyberspace as an environment in which they can pursue their strategic objectives free of the constraints posed by physical borders—the Great Internet Wall of China is a striking defensive example.

Widely diverging interests also make it difficult to find a common approach: while some advocate the negotiation of an international instrument to protect the free use of the Internet, others are opposed as they fear that such an instrument could lead to more illiberal control of the Internet.[55]

Furthermore, the dual use character of goods poses specific challenges: while software to chase child pornography on the Internet would be welcome, its use can be diverted easily to control the Internet in general and to suppress the freedom of expression.

The EU's commitment to rules-setting was confirmed in the 2019 policy paper on "The EU action to strengthen rules-based multilateralism". Enhancing cybersecurity is one of the three strands for EU action to build an effective multilateral system in extending "multilateralism to new global

[54] Forum for Cooperation on the Indopacific, Paris, 22 February 2022; at https://presidence-francaise.consilium.europa.eu/en/news/press-release-forum-for-cooperation-in-the-indopacific/.

[55] See **Cai** in this volume.

realities" in exercising leadership which draws on the "EU's normative capacity, autonomy and influence".

Implementing data protection[56] rules and measures to enhance cybersecurity allow the EU to act as standard setter also in the area of transformative technology such as AI while ensuring the respect of values such as human rights, ethical and legal standards and safeguards. The High Representative/Vice President (HRVP) specifically linked multilateralism and the rules-based order in cyberspace, lending the EU's support to fighting malicious cyber attacks and to the application of international law to cyber as discussed in the UN.

In line with the 2021 policy paper on "strengthening the EU's contribution to rules-based multilateralism"[57] the need for international cooperation to fight cybercrime, for extending multilateralism to new global issues in shaping global norms and standards was confirmed. "Digital taxation, data protection and privacy, tackling disinformation, illegal content online, 5G, internet governance, cybersecurity, digital finance including payments and cryptocurrencies, and e-government" are defined as areas where rules are insufficient, therefore requiring EU priority engagement. Thereby, the EU is conscious of the "need to strike a delicate balance between pursuing the need for technological sovereignty while upholding the openness of the internet and fundamental rights".[58] A commitment to supporting "the development of normative frameworks for new weapons technologies and new dimensions of warfare such as cyber and outer space"[59] adds an ethical dimension to the political delicacy.

6 Conclusions: Upgrading Cyber Diplomacy

Drawing in particular on the 2020 new Cyber Security Strategy which aims at establishing an operational capacity to prevent, deter and respond to cyber threats as well as the recent additional elements like the post-pandemic digital and green transformation, the analysis of this chapter shows that cyber diplomacy has its own characteristics.

[56] See **Christou/Lee** in this volume.

[57] EC – Joint Communication to the European Parliament and the Council, *On strengthening the EU's contribution to rules-based multilateralism* (Brussels: February 17, 2021), https://www.europarl.europa.eu/meetdocs/2014_2019/plmrep/COMMITTEES/AFET/DV/2021/04-12/COM_JOIN20210003_EN.pdf.

[58] Ibid., 8.

[59] Ibid., 10.

It fulfils three important functions: first, supporting externally the relevant internal EU policies; thereby providing, secondly, a protective as well as, thirdly, an offensive tool for the EU's Cyber Diplomacy Tool Box.

The first category, support, includes spreading relevant EU standards (data protection, privacy, encryption, etc.) to establish a level playing field in promoting EU standards such as the GDPR. Preparing the stage for standard setting and for policing the implementation of commitments made are crucial tools. Serving policymakers in collecting intelligence of legal and technological developments through the network of EU Delegations adds to soft power. This also raises awareness for international developments in the field of cyber, including threat scenarios through reporting on cyber attacks and their modalities worldwide. Success depends on the involvement of the private sector as tech companies play a particularly important role in technological progress, are part of the critical infrastructure and play a crucial role in the development of norms and standards. The EU and the US joining in 2021 the 2018 Paris Call for Trust and Security in Cyberspace[60] is an important step in achieving public-private partnerships also through cyber-related Track 1.5 and Track 2 diplomacy.[61]

To this "all-in" approach the European institutions and member states need to be associated in order to stay on the curve of technological development or ahead of it. This is a significant shortcoming in the development of cyber policymaking and diplomacy as member states are reluctant to share resources because of varying threat perceptions, technological differences and concerns over competition. The Strategic Compass takes up this important aspect in developing a common European threat scenario and security culture.

The second protective role of cyber diplomacy includes assuring the security of personal data of citizens, working for a trustworthy IT environment, analyzing intelligence, unmasking disinformation, developing an early warning system for cyber attacks and detecting the contamination of systems and networks. Capacity building to jointly meet all kinds of malicious cyber attacks is necessary. This requires the development of measures to secure critical infrastructure, supply and production lines, trade and

[60] Paris Call For Trust and Security in Cyberspace, https://pariscall.international/en/ (accessed 13 November 2021).

[61] Camino Kavanagh et al (2021). Quiet Conversations: Observations from a decade of practice in cyber-related track 1.5 and track 2 diplomacy. EU Cyber Direct, November 2021; https://eucd.s3.eu-central-1.amazonaws.com/eucd/assets/Nh3fdYmX/quiet-conversations-16-11-21.pdf.

financial systems, transport lines, space and so on. In terms of policy making the dependence on a single provider of essential technologies or hardware has to be avoided. Alternatives need to be built through diversification.

Establishing and maintaining a level playing field for fair competition is part of the protective function. Again, the main challenge is China which has made increasing its digital influence the goal of its policy and diplomacy (self-sufficiency in high-tech industries by 2025 and overtaking competitors by 2049; as spelt out in the 2021 Outline for National Standardisation Development[62] on top of the 2035 China Standards Project). In terms of diplomacy this means countering efforts to promote Chinese standards through the Digital Silk Road as well as through standard dispersing through development assistance and commercial success, such as low-cost Huawei iPhones. Positively, the EU needs to start quickly talks with Japan, South Korea and Singapore on deeper cooperation on data flows, data-based innovation and allowing more digital trade. India is an important partner on emerging technologies. In short, digital diplomacy needs to be mainstreamed.

Another essential feature of the protective function of diplomacy is international support in the handling of frequent low-threshold attacks. Economic operators and individuals need protection against blackmailing and other cybercrimes including cyber-bullying. Contingency plans to meet worst-case scenarios—for example, the destruction or paralysation of critical infrastructure—have to be drawn up in cooperation with like-minded partners. In addition, diplomacy is challenged to negotiate international agreements to promote EU standards and to assure their implementation.

The third offensive part includes taking effective countermeasures such as sanctions and other countermeasures in response to cyber attacks.[63] A multilateral framework would increase effectiveness and make the EU a sought-for partner for countries like South Korea, which is hesitant for geopolitical reasons to act alone. The EU can leverage access to the EU Single Market. Strategic autonomy also requires the ability to block the

[62] Ministry of Justice of the People's Republic of China, *China issues outline to promote standardized national development*, 11 October 2021; http://en.moj.gov.cn/2021-10/11/c_666741.htm.

[63] The Strategic Compass presented in November 2021 foresees measures for cyber deterrence in 2022.

exterritorial application of national laws of third countries. The spreading of standards increases the competitiveness of businesses. Thus, working with and through standard setting organisations and associations like the International Telecommunications Union or the International Standard Organisation is a crucial offensive tool of a protective nature. This also shows the need for public-private cooperation as the private sector plays an essential role in this area.

Realising strategic openness while protecting EU values, "EU cyber diplomacy should be made more coherent in its supranational, democratic, and economic/technological dimension".[64] Winning the battle for digital governance is essential as "whoever wins the global debate over the rules, standards, and norms that govern data privacy and data flows, technology standards, cybersecurity, and critical technologies will have a major competitive advantage in the economy of 2030".[65]

The EU has laid the foundations but needs to urgently make further progress as One also referred to as Team Europe. As diplomacy starts at home, only implementing agreed strategies systematically and not piecemeal whenever a majority is reached will allow the EU to act globally as one in the form of a comprehensive and joint EU cyber diplomacy which can take care of European interests and values.

References

ASEAN. *ASEAN-EU Statement on Cybersecurity Cooperation*. Thailand: 2019, https://asean.org/storage/2019/08/ASEAN-EU-Statement-on-Cybersecurity-Cooperation-FINAL.pdf.

BBC News. "NATO: Cyber-attack on one nation is attack on all." *BBC News*, August 27, 2019. https://www.bbc.com/news/technology-49488614.

Bendiek, Annegret. "The EU as a Force for Peace in International Cyber Diplomacy." *SWP Comment* no. 19 (April 2018). https://www.swp-berlin.org/fileadmin/contents/products/comments/2018C19_bdk.pdf.

[64] Annegret Bendiek and Matthias Kettemann, "Revisiting the EU Cybersecurity Strategy: A Call for EU Cyber Diplomacy," *SWP Comment* no. 16 (February 2021), https://www.swp-berlin.org/10.18449/2021C16/.

[65] Matthew P. Goodman, "Time to Align on Digital Governance," Commentary, Center for Strategic & International Studies, January 24, 2020, https://www.csis.org/analysis/time-align-digital-governance.

Bendiek, Annegret, Kettemann Matthias. "Revisiting the EU Cybersecurity Strategy: A Call for EU Cyber Diplomacy," SWP Comment no. 16 (February 2021), https://www.swp-berlin.org/10.18449/2021C16/.

Council of the European Union. *Council Conclusions on Cyber Diplomacy*. Brussels: February 15, 2015. http://data.consilium.europa.eu/doc/document/ST-6122-2015-INIT/en/pdf.

Council of the European Union. *Decision establishing the list of projects to be developed under PESCO*. Brussels: March 1, 2018a. http://data.consilium.europa.eu/doc/document/ST-6393-2018-INIT/en/pdf.

Council of the European Union. *Enhanced EU Security Cooperation in and with Asia – Council conclusion (9265/1/18 REV 1)*. Brussels: May 28, 2018b. http://www.consilium.europa.eu/media/35456/st09265-re01-en18.pdf.

Council of the European Union. *EU Cyber Defence Policy Framework (2018 update)*. Brussels: November 19, 2018c. https://data.consilium.europa.eu/doc/document/ST-14413-2018-INIT/en/pdf.

Council of the European Union. "Cyber-attacks: Council is now able to impose sanctions." Press Releases, May 17, 2019a. https://www.consilium.europa.eu/en/press/press-releases/2019/05/17/cyber-attacks-council-is-now-able-to-impose-sanctions/.

Council of the European Union. "EU imposes the first ever sanctions against cyber-attacks," Press release, July 30, 2020a, https://www.consilium.europa.eu/en/press/press-releases/2020/07/30/eu-imposes-the-first-ever-sanctions-against-cyber-attacks/

Council of the European Union. *Draft Council conclusions on the EU's Cybersecurity Strategy for the Digital Decade*. Brussels: March 9, 2021a. https://data.consilium.europa.eu/doc/document/ST-6722-2021-INIT/en/pdf.

Council of the European Union. Draft Council conclusions on the EU's Cybersecurity Strategy for the Digital Decade. 9 March 2021.

Council of the European Union. *EU Strategy for cooperation in the Indo-Pacific*. Brussels: April 16, 2021b. https://data.consilium.europa.eu/doc/document/ST-7914-2021-INIT/en/pdf.

Council of the European Union. *Joint press release: Republic of Korea – EU Leaders' video conference meeting*, 30 June 2020b, accessed 7 October 2020, https://www.consilium.europa.eu/en/press/press-releases/2020/06/30/joint-press-release-republic-of-korea-eu-leaders-video-conference-meeting/.

Council of the European Union, Newsroom, July 12, 2021c, https://newsroom.consilium.europa.eu/events/20210712-foreign-affairs-council-july-2021.

DiploFoundation. "Digital diplomacy | E-diplomacy | Cyber diplomacy." Accessed February 2, 2020. https://www.diplomacy.edu/e-diplomacy.

EU Council, Draft Council Conclusions on a Framework for a Joint EU Diplomatic Response to Malicious Cyber Activities ("Cyber Diplomacy Toolbox")

(Brussels: June 7, 2017) http://data.consilium.europa.eu/doc/document/ST-9916-2017-INIT/en/pdf.

EURAXESS, Press Release 12 November 2021, https://euraxess.ec.europa.eu/worldwide/south-korea/commission-invest-nearly-%E2%82%AC2-billion-digital-europe-programme-advance.

European Commission, "Commission Recommendation of 26.3.2019 Cybersecurity of 5G networks," Library, March 26, 2019a, https://digital-strategy.ec.europa.eu/en/library/cybersecurity-5g-networks.

European Commission. "Code of Practice against disinformation: Commission recognises platforms' efforts ahead of the European elections." Press Corner, May 17, 2019b. https://europa.eu/rapid/press-release_STATEMENT-19-2570_en.htm.

EU Council, Draft Council Conclusions on a Framework for a Joint EU Diplomatic Response to Malicious Cyber Activities ("Cyber Diplomacy Toolbox") (Brussels: June 7, 2017), http://data.consilium.europa.eu/doc/document/ST-9916-2017-INIT/en/pdf.

EU Council, Draft Council conclusions on the EU's Cybersecurity Strategy for the Digital Decade (Brussels, 9 March 2021), https://data.consilium.europa.eu/doc/document/ST-6722-2021-INIT/en/pdf.

EC-Joint Communication

EC – Joint Communication to the European Parliament and the Council, On strengthen- ing the EU's contribution to rulesbased multilateralism (Brussels: February 17, 2021), https://www.europarl.europa.eu/meetdocs/2014_2019/plmrep/COMMITTEES/AFET/DV/2021/04-12/COM_JOIN20210003_EN.pdf.

EC-Joint Communication to the European Council and European Parliament, *The EU strategy for cooperation in the Indo-Pacific* (Brussels, September 16, 2021) https://eeas.europa.eu/headquarters/headquarters-homepage/104126/joint-communication-indo-pacific_en

European Commission, Communication from the Commission to the European Parliament, the European Council, the Council, the European economic and social Committee and the Committee of the regions – on the EU Security Union Strategy (Brussels: July 24, 2020a), https://eur-lex.europa.eu/legal-content/EN/TXT/PDF/?uri=CELEX:52020DC0605&from=EN.

European Commission. *Proposal for a Directive of the European Parliament and of the Council on the resilience of critical entities.* Brussels: December 16, 2020b. https://ec.europa.eu/home-affairs/sites/default/files/pdf/15122020_proposal_directive_resilience_critical_entities_com-2020-829_en.pdf.

European Commission. "EU Cybersecurity: Commission proposes a Joint Cyber Unit to step up response to large-scale security incidents." Press release, June 23, 2021a. https://ec.europa.eu/commission/presscorner/detail/en/IP_21_3088.

European Commission. Commission work programme 2022 – Making Europe stronger together, (Brussels, 19 October 2021b), https://ec.europa.eu/info/sites/default/files/com2021_645_en.pdf

European Commission and HR/VP contribution to the European Council. "EU-China – A strategic outlook." March 12, 2019, https://ec.europa.eu/info/sites/default/files/communication-eu-china-a-strategic-outlook.pdf.

European Commission and High Representative of the Union for Foreign Affairs and Security Policy, The EU strategy for cooperation in the Indo-Pacific (Brussels: September 21, 2021), 15, https://ec.europa.eu/info/sites/default/files/jointcommunication_indo_pacific_en.pdf

European Commission – Joint Communication to the European Parliament, Council, European Economic and Social Committee and Committee of Regions. *Cybersecurity Strategy of the European Union: An Open, Safe and Secure Cyberspace.* Brussels: February 2, 2013. https://eeas.europa.eu/archives/docs/policies/eu-cyber-security/cybsec_comm_en.pdf.

European Commission – Joint Communication to the European Parliament, Council, European Economic and Social Committee, Committee of the Regions and European Investment Bank. *The EU Strategy Connecting Europe and Asia – Building blocks for an EU Strategy.* Brussels: September 19, 2018. https://eeas.europa.eu/sites/eeas/files/joint_communication_-_connecting_europe_and_asia_-_building_blocks_for_an_eu_strategy_2018-09-19.pdf.

European Commission. "Recovery Plan for Europe," accessed August 10, 2021c, https://ec.europa.eu/info/strategy/recovery-plan-europe_en.

European Commission. Global digital partnerships; at https://ec.europa.eu/international-partnerships/topics/digital-partnerships_en (accessed 26 February 2022a).

European Commission. "Joint Statement: EU and Singapore agree to strengthen bilateral partnership on digital trade". Press release, 7 December 2021d; at https://trade.ec.europa.eu/doclib/press/index.cfm?id=2337.

European Political Strategy Centre – now known as IDEA (European Commission). *Rethinking Strategic Autonomy in the Digital Age.* EPSC Strategic Notes no. 30, July 2019. https://op.europa.eu/en/publication-detail/-/publication/889dd7b7-0cde-11ea-8c1f-01aa75ed71a1/language-en/format-PDF/source-118064052.

EEAS. Questions and Answers about the East StratCom Task Force News stories (Brussels 28 April 2021), https://eeas.europa.eu/headquarters/headquarters-homepage/2116/-questions-and-answers-about-the-east-stratcom-task-force_en

Fiott, Daniel. "Strategic autonomy: towards 'European sovereignty' in defence?." *EUISS Brief* 12 (2018). https://www.iss.europa.eu/content/strategic-autonomy-towards-%E2%80%98european-sovereignty%E2%80%99-defence.

Fiott, Daniel. *The Yearbook of European Security 2019.* Paris: EUISS, 2019. https://www.iss.europa.eu/sites/default/files/EUISSFiles/yes2019.pdf.

French Presidency of the European Council. Forum for Cooperation on the Indopacific, Paris, 22 February 2022; at https://presidence-francaise.consilium.europa.eu/en/news/press-release-forum-for-cooperation-in-the-indopacific/.

GIZ, *Enhancing Security Cooperation in and with Asia*, accessed 28 October 2021, https://www.giz.de/en/worldwide/87412.html.

Goodman, Matthew P. "Time to Align on Digital Governance." Commentary, Center for Strategic & International Studies (CSIS), January 24, 2020. https://www.csis.org/analysis/time-align-digital-governance.

Joint Communication to the European Parliament, the Council, the European Economic and Social Committee, the Committee of the Regions and the European Investment Bank (2021). *The Global Gateway.* JOIN (2021) 30 final, 1 December 2021; at https://ec.europa.eu/info/sites/default/files/joint_communication_global_gateway.pdf.

Camino Kavanagh, Madeline Carr, Nils Berglund (2021). Quiet Conversations: Observations from a decade of practice in cyber-related track 1.5 and track 2 diplomacy. EU Cyber Direct, November 2021; https://eucd.s3.eu-central-1.amazonaws.com/eucd/assets/Nh3fdYmX/quiet-conversations-16-11-21.pdf.

Lété, Bruno. "Cooperation in cyberspace." In *The EU and NATO: the essential partners,* edited by Gustav Lindstrom and Thierry Tardy, 28–36. Paris: European Union Institute for Security Studies (EUISS), 2019.

Ministry of Justice of the People's Republic of China, *China issues outline to promote standardized national development*, 11 October 2021; http://en.moj.gov.cn/2021-10/11/c_666741.htm.

Ministry of Trade and Industry (MTI) of Singapore. "Digital Economy Partnership (DEPA)"; at https://www.mti.gov.sg/Improving-Trade/Digital-Economy-Agreements/The-Digital-Economy-Partnership-Agreement (accessed 26 February 2022).

NIS Cooperation Group. *Cybersecurity of 5G networks EU Toolbox of risk mitigating measures.* CG Publication, January 29, 2020. https://ec.europa.eu/digital-single-market/en/news/cybersecurity-5g-networks-eu-toolbox-risk-mitigating-measures.

Paris Call For trust and security in cyberspace, https://pariscall.international/en/ (accessed 13 November 2021).

Riordan, Shaun, and Mario Torres Jarrín. *Global Policy Perspective Report: Cyberdiplomacy.* Salamanca-Stockholm: European Institute of International Studies, 2020. http://www.ieeiweb.eu/wp-content/uploads/2020/01/Cyberdiplomacy_enero.pdf.

Scheidt, Mélanie. "The European Union versus External Disinformation Campaigns in the Midst of Information Warfare: Ready for the Battle?." *EU Diplomacy Papers* no 1 (August 2019). https://www.coleurope.eu/study/eu-international-relations-and-diplomacy-studies/research-publications/eu-diplomacy-papers.

The European Centre of Excellence for Countering Hybrid Threats (HybridCoE). "Home". Accessed July 21, 2021, https://www.hybridcoe.fi/.

EU-South Korea Cooperation on Cybersecurity, Data Protection and Emerging Technologies

George Christou and Ji Soo Lee

1 Introduction

The increased and accelerated digitalisation of society has elevated the issue of how a safe, secure and resilient infrastructure and ecosystem can be constructed so that threats can be minimised and opportunities maximised. Building and sustaining a trustworthy digital environment and tools, however, requires a multi-stakeholder approach and global coordination and cooperation; the non-territorial nature of cyberspace requires that states, regional organisations and corporations work together towards creating a global cyber ecosystem that delivers sustainable, secure and dynamic digital growth. This has become even more urgent given the

G. Christou (✉)
University of Warwick, Coventry, UK
e-mail: G.Christou@warwick.ac.uk

J. S. Lee
Hankuk University of Foreign Studies, Seoul, South Korea

© The Author(s), under exclusive license to Springer Nature Switzerland AG 2022

G. Boulet et al. (eds.), *Cybersecurity Policy in the EU and South Korea from Consultation to Action*, New Security Challenges,
https://doi.org/10.1007/978-3-031-08384-6_3

increased digitalisation of life triggered by the covid-19 pandemic and, with this, the increased vulnerability of all to cyber attacks. The geopolitics of cyberspace and the clash of ideas over how the Internet, emerging technologies and cyberspace should be governed have also served to add to the need for cooperation between actors that seek to maintain a freely accessible Internet that does not threaten the rule of law, fundamental rights and freedom and democracy.

Focusing on the EU and South Korea, considered normatively like-minded in the cyber sphere, this chapter will hone in on how cooperation has evolved to address shared threats and common concerns and interests relating to cybersecurity, data protection and emerging technologies. Both the EU[1] and South Korea[2] place an important emphasis on international collaboration and coordination as a means to addressing the threats emanating from the cyber sphere.[3] At the centre of EU-South Korea cooperation on cyber issues is the cyber dialogue that was launched in November 2013 with the express aim of strengthening collaboration and cooperation bilaterally and multilaterally on a range of issues that impact global cybersecurity, stability and resilience.[4] The commitment to launch a High Level Dialogue on digital economy has also emphasised the need for future cooperation not just on a rules-based order for cyberspace but also on issues such as disinformation, "responding to the Fourth Industrial Revolution, harnessing human-centric digitalisation and promoting safe and ethical implementation of Artificial Intelligence technologies".[5] More concretely, in facilitating the creation of a trustworthy digital environment for trade through safe and secure cross-border data flows, the EU and

[1] European Commission and High Representative of the Union for Foreign Affairs and Security Policy (2020), Joint Communication to the European Parliament and the Council, "The EU's Cybersecurity Strategy for the Digital Decade", Brussels, 16.12.2020 JOIN (2020) 16 final, 19.

[2] National Security Strategy, chapter VII.2/Strengthening the capacity to address cybersecurity threat, December 2018.

[3] See also Hannes Ebert and Laura Groenendaal, "Cyber Resilience and Diplomacy in the Republic of Korea: Prospects for EU Cooperation", Digital Dialogue, EU Cyber Direct (August 2020), 3.

[4] George Christou and Ji Soo Lee, "EU–ROK cooperation on cyber-security and data protection" in *EU-Korea Security Relations*, ed. Nicola Casarini (London: Routledge, 2021), 55.

[5] European Commission, "Joint press release: Republic of Korea -EU Leaders' video conference meeting", June 30, 2020. https://ec.europa.eu/commission/presscorner/detail/en/IP_20_1226.

South Korea have made significant progress with regard to the convergence of their data protection rules through the data adequacy decision of 17 December 2021 that will ensure high levels of protection with regard to privacy and personal data.

More broadly, South Korea is one of the EU's strategic partners and is a country with whom cooperation is viewed as critically important in addressing the challenges of the twenty-first century in a post-covid world.[6] Indeed, South Korea is the only country to have effective agreements with the EU in the economic (Free Trade Agreement), political (Framework Agreement) and security (Framework Participation Agreement for EU crisis management operations) spheres, making it for some scholars one of the EU's closest international partners.[7] In this context then, it will be argued in this chapter that the EU and South Korea share the same global vision and a common interest in ensuring the construction of a rules-based order for cyberspace. They are increasingly convergent on data protection rules and the ethics of new technologies, but divergences exist, and cooperation can be significantly improved beyond dialogue, information sharing and trust-building.

This chapter proceeds in the following way. First it sketches out the respective approaches of the EU and South Korea in relation to cybersecurity, data protection and emerging technologies. The second section moves to discuss common threats and interests in these spheres and to chart cooperation between the EU at bilateral, multilateral and regional levels. Finally, we offer thoughts in the concluding section on opportunities and potential obstacles to future cooperation between the EU and South Korea on issues of cybersecurity, data protection and emerging technologies.

[6] Michael Reiterer, "The 10th anniversary of the EU-Korea Strategic partnership", VUB-KF Chair, Policy Brief, Issue 2020/09, July 2020. https://www.korea-chair.eu/wp-content/uploads/2020/07/KFVUB_Policy-Brief-2020-09.pdf.

[7] Brigitte Dekker and Maaike Okano-Heijmans, "The Implications of China's Techno-Nationalism for the European Union and the Republic of Korea, Europe and the World Articles, Clingendael Magazine, 27 October 2020. https://www.clingendael.org/publication/eu-republic-korea-digital-connectivity-unitedwe-must-stand.

2 The EU's and South Korea's Approaches to Cybersecurity, Data Protection and Emerging Technologies

2.1 *The European Union*

2.1.1 Cybersecurity

The EU's approach to cybersecurity has evolved incrementally since the 1970s. We can trace a move from a soft law approach to a more formal hard law approach,[8] or genesis to formalisation—and with this continuity and change in guiding principles, programmes and practice.[9] It is also evident that certain triggers, such as the distributed denial-of-service (DDoS) attacks on Estonian public and private institutions and infrastructure in 2007, have provided critical moments for the EU and NATO to reflect on how they would address threats emanating from cyberspace.[10]

After 2007 various cybersecurity initiatives were developed within the context of the Digital Agenda for Europe (2010) and the EU's Internal Security Strategy (2010)[11] alongside guiding (normative) principles for Internet governance (2011),[12] with the first EU Cybersecurity Strategy[13] published in 2013. This Strategy integrated the previously separate mandates on cybersecurity,[14] setting out priorities in relation to Network and

[8] Elaine Fahey "The EU's cybercrime and cyber-security rulemaking: mapping the internal and external dimensions of EU security", *European Journal of Risk Regulation*, 5, Issue 1 (2014), 46–60.

[9] Helena Carrapico and Benjamin Farrand, "Discursive continuity and change in the time of Covid-19: the case of EU cybersecurity policy", *Journal of European Integration*, 42, Issue 8 (2020), 1111–1126.

[10] George Christou "The Collective Securitisation of EU Cybersecurity", *West European Politics*, 42, Issue 2 (2019), 278–301.

[11] European Commission "A Digital Agenda for Europe" (No.COM (2010)245 Final/2), Brussels; European Commission "The EU internal Security Strategy in Action: Five Steps Towards a More Secure Europe" (No. COM (2010) 673 Final), Brussels.

[12] European Commission "European principles and guidelines for Internet resilience and stability", March 2011. http://ccpic.mai.gov.ro/docs/guidelines_internet_fin.pdf.

[13] European Commission and High Representative of the European Union for Foreign Affairs and Security Policy/Vice-President of the Commission, "Cybersecurity Strategy of the EU: An Open, Safe and Secure Cyberspace", Brussels, 7 February 2013.

[14] George Christou "Cybersecurity in the European Union: Resilience and Adaptability in Governance Policy" (Houndmills, Basingstoke: Palgrave Macmillan, 2016).

Information Security; cybercrime and cyber attacks; cyber defence; developing the industrial and technological resources for cybersecurity; and, pertinent to the argument in this chapter, establishing a coherent international cyberspace policy for the EU in order to promote core EU values. At the core of this first Strategy was the 2013 Network and Information Systems (NIS) Directive[15] which was implemented by EU member states in May 2018 and has been revised, following review, to make it more flexible and future-proof in the context of new and emerging technologies.[16]

Since 2013 the EU has deepened and widened its approach to cybersecurity reflecting the increased urgency within which the dangers of cyber threats and risk are perceived within the EU. Persistent everyday cyber breaches, including against EU institutions, have increased threat perceptions and incentivised the EU to enhance significantly its ability to deal with cyber attacks.[17] The covid-19 pandemic which brought with it an acceleration in societal digitisation across all areas of life also witnessed an increase in disinformation as well as cybercrime and cyber attacks as cyber criminals sought to exploit and take advantage of increased online activity. The pandemic, then, injected further urgency within EU policymaking circles to strengthen online safety and resilience so that it could realise its "Digital Future" and preserve its political vision and model of cyberspace.[18]

[15] On cybercrime, the main piece of legislation was the 2013 Directive on Attacks against Information Systems, which member states had to transpose into domestic law by 2015. The Directive (EU) 2016/1148 was introduced to further bolster and ensure enhanced the security of network and information systems within the European Union.

[16] It is now closely aligned with the proposal for the Critical Entities Resilience Directive which enhances the NIS 2 Directive by imposing new security requirements on to companies that provide essential services as well as sector-specific Regulations such as the Digital Operational Resilience Act (DORA) for the financial sector.

[17] ENISA, Threat Landscape Report, 2020. https://www.enisa.europa.eu/topics/threat-risk-management/threats-and-trends?tab=details EUROPOL, Internet Organised Crime Threat Assessment, 2020.

[18] See Carrapico and Farrand, "Discursive continuity". https://www.europol.europa.eu/activities-services/main-reports/internet-organised-crime-threat-assessment-iocta-2020.

The increasing strategic importance of cybersecurity has been evidenced through its prioritisation in EU security policy documents,[19] as well as its digital growth, industrial, data and disinformation strategies and action plans.[20] In terms of policy and practice, the EU has developed a number of initiatives, tools, mechanisms and processes—regulatory and operational—across its multiple cyber mandates in order to address cybersecurity risks and threats. The EU has also articulated specific cyber diplomacy goals[21] and developed a framework for joint EU diplomatic response to malicious cyber activities and its cyber sanctions regime[22] (see chapter by Michael **Reiterer**). The EU's Cyber Security Act[23]—part of a broader set of measures introduced in its Cybersecurity Package in 2017[24]—as well as the EU's 'Cybersecurity Strategy for the Digital Decade' is also important. They seek to—alongside other digital technology and disinformation initiatives[25]—enhance vertical and horizontal coherence and strengthen the EU's cybersecurity capability and resilience through its ambition to

[19] European Commission. "European agenda on security", 28 April, COM (2015) 185 final. European Commission and European External Action Service. Joint Communication to the European Parliament and the Council, Joint Framework on Countering Hybrid Threats: A European Union response. 6 April, JOIN (2016) 18 final. Council of the European Union "Shared Vision, Common Action: A Stronger Europe. A Global Strategy for the EU's Foreign and Security Policy" (2016). European Commission, Communication from the Commission to the European Parliament, The European Council, The Council, The European Economic and Social Committee and the Committee of the Regions on the "EU Security Union Strategy", Brussels, 24.7.2020, COM (2020) 605 final.

[20] Council of the European Union. "Council Conclusions on Shaping Europe's Digital Future (No. OJ 2020C: 202); Council of the European Union. "Council Conclusions on Covid-19 (No." In OJ C, 57/4). European Commission. "Shaping Europe's Digital Future"; European Commission. "A European Strategy for Data" (No. COM (2020) 66); European Commission. "A New Industrial Strategy for Europe" (No. COM (2020) 102 F).

[21] Council of Ministers, 'Council Conclusions on Cyber Diplomacy', 6122/15, Brussels, 11 February 2015.

[22] Council Decision 2019/797; Council Regulation 2019/796.

[23] Regulation (EU) 2019/881 of the European Parliament and of the Council of 17 April 2019 on ENISA (the European Union Agency for Cybersecurity) and on information and communications technology cybersecurity certification and repealing Regulation (EU) No 526/2013 (Cybersecurity Act) (Text with EEA relevance) PE/86/2018/REV/1.

[24] European Commission and High Representative of the Union for Foreign Affairs and Security Policy. Joint communication to the European Parliament and the Council, Resilience, Deterrence and Defence: Building strong cybersecurity for the EU, JOIN/2017/0450 final.

[25] For example, the EU's proposed Digital Services Act and Digital Markets Act.

become technologically sovereign and strategically autonomous in cyberspace.

Finally, the EU's normative and governance position on cybersecurity has included support for multi-stakeholder solutions, and a global, stable, open, accessible Internet "based on fundamental rights and freedoms as enshrined in the Charter of Fundamental Rights of the European Union".[26] In addition, the EU is aware that global disagreement and contestation on many aspects of cybersecurity, from norms to standards or appropriate legal conventions for fighting cybercrime (see chapters by Tatiana **Tropina** and Gibum **Kim**), can undermine any attempt to create a secure cyberspace for all. The EU has thus been consistently clear in its position that cooperation is essential with international partners to create global principles compatible with EU values and interests.

2.1.2 Data Protection

Developing a robust legal framework for the protection of citizens' data has been integral to the EU's approach for fostering a trustworthy online environment for EU consumers; an issue which has become even more important in the covid-19 era. The first comprehensive piece of legislation published in 1995 was the Data Protection Directive (DPD), which sought to protect the fundamental rights of individuals and ensure there was no impediment to the free flow of personal data. Following review of the DPD in 2003 and 2007 and extensive consultation by the Commission, a proposal was put forward in 2012 for a new General Data Protection Regulation (GDPR). Approved by the European Parliament in 2016 after much internal contestation, the GDPR was implemented by the EU member states in 2018, a key aim of which was "to build a stronger and more coherent data protection framework in the EU".[27] It built on existing DPD principles, with legislative proposals in the form of a Regulation for protecting individuals in relation to processing and free movement of personal data and a Directive to protect individuals in relation to processing of personal data flows in police and criminal matters. The European Commission also proposed the e-privacy Regulation[28] in January 2017

[26] European Commission and High Representative 2013, 4.

[27] European Commission. "Proposal for a regulation of the European Parliament and of the Council on the protection of individuals with regard to the processing of personal data and on the free movement of such data (General Data Protection Regulation)", COM (2012) 11 final, Brussels, 25.1.2012, 2.

[28] Due to delays, however, the Regulation is not likely to take effect until 2022.

which was intended to complement the general rules in the GDPR by providing specific rules governing electronic communications.

Such reform of the EU's data protection framework was seen as essential in order to: first, restore public trust in using online platforms seen as essential for achieving the EU's Digital Agenda for Europe;[29] second, providing a more robust legislative protection for Europeans regarding privacy and data protection rights following the large-scale collection of personal data through US surveillance programmes that was exposed in the Snowden disclosures in June 2013. To this end, there was also an increase in judicial activism, with the Court of Justice of the EU annulling the Data Retention Directive (2006),[30] which was seen to infringe the right to privacy; and invalidating the EU-US Safe Harbour Agreement (in 2014) and its replacement the EU-US Privacy Shield (in 2020) as they were not seen to provide sufficient protection for European citizens in the transfer of their personal data to the US.

The GDPR, for some, is reflective of the EU's normative and regulatory power as well as its ambition to be technologically sovereign (see chapter by Anna-Maria **Osula**). It also demonstrates the EU's ability to influence other countries when it comes to standards for privacy and data protection; in particular countries such as Japan or South Korea which want to strengthen commercial ties with the EU. This said, the GDPR has also been criticised for precisely the same reason—that is, that the 'Brussels effect' can marginalise many countries that cannot adopt EU standards due to the burden this puts on them in terms of cost and administration. The GDPR data protection and privacy provisions have also caused considerable rifts in the transatlantic context—in terms of both the commercial and criminal dimensions (see chapter by Tatiana **Tropina**) when it comes to the treatment of personal data and upholding and protecting the rights of EU citizens.

2.1.3 *Emerging Technologies*

The EU approach to emerging technologies is underpinned by the normative approach for cybersecurity and a clear ambition to be a global

[29] European Commission. Communication from the Commission to the European Parliament, the Council, the European Economic and Social Committee and the Committee of the Regions, "A Digital Agenda for Europe", COM (2010) 245, 19 May 2010.

[30] Directive 2006/24/EC regulated data retention where data was generated or processed in connection with the provision of publicly available electronic communications services or of public communications networks.

regulatory leader and standard setter in relation to artificial intelligence (AI) and 5G.[31] In seeking to make "Europe fit for the Digital Age" the Commission recognises that it must accelerate its digital transformation and strengthen its leadership in order to ensure that the potential benefits of newly emerging technologies can be leveraged across multiple sectors of economy and society.[32]

The EU's policy on AI has a strong legal, normative and ethical dimension, which takes a risked-based approach that seeks to ensure that mandatory legal requirements built on strong ethical guidelines can be imposed on AI systems that are defined as high risk.[33] The AI Act is the EU's main regulatory tool and is "guided by the EU's underlying idea that developing trustworthy technologies will prompt [through international cooperation] the global uptake of AI in Europe". Thus, the EU has sought to balance the preservation of individual safety and fundamental rights with ensuring that innovation in AI is not stifled. This said, some have suggested that such a risk-based regulatory approach might not be compatible with those of key strategic partners such as the US, which likes to take a more hands-off approach to technology and that takes the view that regulation can inhibit innovation.[34]

On 5G, the EU's Cybersecurity Strategy states clearly that it should be underpinned by the highest security standards so that EU citizens and companies can maximise its benefits going forward. Similar to AI the EU has, through the EU 5G Toolbox, proposed a comprehensive risk-based approach to 5G cybersecurity which consists of three interrelated objectives: achieving convergence in the risk mitigation strategies of EU member states, supporting capacity building and exchange of knowledge, and promoting EU strategic security objectives, in particular supply chain

[31] European Commission and High Representative 2020, 8.

[32] Margrethe Vestager, Executive Vice-President for a Europe fit for the Digital Age, "Answers to the European Parliament questionnaire to the Commissioner-designate", 4–6. https://ec.europa.eu/commission/commissioners/sites/default/files/commissioner_ep_hearings/answers-ep-questionnaire-vestager.pdf.

[33] Margrethe Vestager, Executive Vice-President for a Europe fit for the Digital Age, "Europe fit for the Digital Age: Commission proposes new rules and actions for excellence and trust in Artificial Intelligence", 2021. https://digital-strategy.ec.europa.eu/en/news/europe-fit-digital-age-commission-proposes-new-rules-and-actions-excellence-and-trust-artificial.

[34] Eve Gaumond, "Artificial intelligence Act: What is the EU Approach for AI?", LAWFARE.

resilience.[35] Such an approach has been developed in order to provide clear guidance on a range of security measures—technical and strategic in nature[36]—that will primarily serve to mitigate risks and ensure the secure deployment of 5G networks across Europe.

However, given the geopolitical challenges that come with 5G technologies, the 5G toolbox is not limited to Europe in terms of its technical or strategic aspects. There is full recognition that whilst the EU is "acting European" it must also think global. Here then, the EU has been strategically open in relation to interest from stakeholders outside the EU on the Toolbox and, indeed, in ensuring that it remains engaged with regard to pushing the merits of its Toolbox approach through its bilateral and multilateral processes. This, in recognition of the fact that providing secure 5G in Europe also involves achieving consensus, or at a minimum, strategic alliances with like-minded states. As important, is the provision of leadership within any such alliances to ensure that any global governance of 5G is driven, underpinned by and compatible with EU values and standards.

2.2 South Korea

2.2.1 Cybersecurity

Cyber attacks have been a driving force behind South Korea's cybersecurity policies (see chapter by Kyu-dok **Hong** and Seong-jong **Song**). Since the development ofNational Cyber Crisis Comprehensive Countermeasures due to a DDoS attack in 2009, South Korea continually developed policies for coping with cyber attacks. Examples include the National Cyber Security Master Plan following a DDoS attack and attack against the NH bank in 2011, National Comprehensive Cyber Security Countermeasures following attacks in 2013 and the National Cyber Security Posture and Capability Strengthening Plan resulting from a nuclear power plant attack in 2015.[37]

The Moon Jae-in Administration's National Security Strategy (NSS) published in December 2018 by the National Security Office (NSO) includes establishing a cooperative system that encompasses public, private and military sectors in order to comprehensively respond to cyber threats that occur regardless of domain and strengthening the NSO's role of the

[35] European Commission and High Representative 2020, 9.
[36] https://digital-strategy.ec.europa.eu/en/library/eu-toolbox-5g-security.
[37] Ebert and Groenendaal, "Cyber Resilience", 8–9.

control tower.[38] It also includes strengthening preventive and response capabilities to cyber threats by increasing the safety of core cyberspace.[39] Furthermore, it recognises the need for international cooperation in all relevant fields such as diplomacy, national defence and information in order to respond to cyber threats which have a transnational nature.[40]

Subsequently, in accordance with the aforementioned NSS and for providing more concrete policy directions for cybersecurity, the NSO published the National Cybersecurity Strategy (NCS) in April 2019. It provides a long-term vision, goals and direction for policies increasing the safety of core national infrastructure, reinforcing cyber threat response capacity, fostering cybersecurity industries and intensifying international cooperation on cybersecurity.[41] The NCS is underpinned by three principles: harmonising cybersecurity with fundamental rights; conducting security activities based on the rule of law and establishing systems for international cooperation and multi-stakeholders' engagement including individuals, businesses and governments.[42] In September 2019, the government of South Korea developed the National Cybersecurity Basic Plan with the aim of facilitating the implementation of the NCS.[43]

2.2.2 Data Protection

South Korea's three major laws on data protection—the Personal Information Protection Act (PIPA), the Act on the Promotion of Information and Communications Network Utilisation and Data Protection (Network Act), and the Credit Information Use and Protection Act (Credit Information Act)—were amended in 2020. The amendment reformed the system for governing data protection by unifying provisions on data protection that had been dispersed in the laws (e.g. the Network Act) into the PIPA, transferring supervisory function for data protection (which was previously in other agencies such as the Ministry of the Interior and Safety, and the Korea Communications Commission) to the Personal Information Protection Commission (PIPC) and upgrading the PIPC to

[38] *National Security Strategy* (National Security Office, 2018), 103–104.

[39] For example, communication networks of public institutions and infrastructures.; *National Security Strategy*, 103–105.

[40] *National Security Strategy*, 103 and 105.

[41] See *National Cybersecurity Strategy* (National Security Office, 2019).

[42] *National Cybersecurity Strategy*, 12.

[43] *National Cybersecurity Basic Plan [국가 사이버안보 기본계획]* (Republic of Korea [관계부처 합동], 2019), 11.

a central administrative agency.[44] The objective of the amendment was to ensure the secure use of personal data for promoting new industries in responding to the Fourth Industrial Revolution and it was also related to discussions on the EU GDPR adequacy decision taken in December 2021.

Under Article 9 of the PIPA, the PIPC is obliged to draw up a Master Plan for personal data protection every three years. In general, the Master Plan contains goals and tasks to be carried forward including improvement of rules, systems and measures to prevent infringement of personal data.[45] Since 2012 the Master Plans had mainly aimed at the stable establishment of the PIPA which was enacted and entered into force in 2011 and the strengthening of each stakeholder's capacities.[46] It was then developed to create an effective protection system and enhance protection activities.[47] The 4th Master Plan (2021–2023), being implemented at the time of writing, reflects the amendment of three major data laws in settling its policy directions.[48] The Plan set up the vision of ensuring a secure and trustworthy digital environment and adopted solid data protection, safe use of personal data and a fair balance between the protection and use as the strategies for realising the vision.[49] To achieve these strategies, detailed tasks (e.g. reinforcing data subject's rights, support for secure use of personal data) were prepared.[50] Importantly, receiving and maintaining adequacy status is included in the Plan as a way to enhance international cooperation.[51] For an effective implementation of the Master Plan, Article 10 of the PIPA requires the heads of central administrative agencies (ministries, the National Intelligence Service, the Board of Audit and

[44] *2021 National Cybersecurity White Paper* (Korean version), (National Intelligence Service et al., May 2021), 21; Christou and Lee, "EU–ROK cooperation on cyber-security and data protection," 68–69.

[45] "Master Plan Implementation Plan: Master Plan", Law & Policy, Personal Information Protection Commission, accessed 17 July 2021. https://www.pipc.go.kr/eng/user/lgp/pbp/mlp/masterPlan.do.

[46] Personal Information Protection Commission, "Master Plan".

[47] See *2020 Annual Report on Personal Information Protection* (Personal Information Protection Commission, 2020), 51. https://www.pipc.go.kr/np/ebook/2020Y_report/index.html (accessed 17 July 2021).

[48] Personal Information Protection Commission, "Master Plan".

[49] Personal Information Protection Commission, "Master Plan".

[50] Personal Information Protection Commission, "Master Plan".

[51] *Master Plan [개인정보 보호 기본계획] (2021–2023)*, Personal Information Protection Commission, (November 2020), 27 and 31. available at https://www.pipc.go.kr/np/default/page.do?mCode=D050010010#LINK.

Inspection, etc.) to establish an implementation plan each year and to carry out the plan subject to deliberation of the PIPC.[52]

2.2.3 Emerging Technologies

Building the foundation to prepare for the Fourth Industrial Revolution by improving regulations, developing core technology and fostering new industries is an important part of 100 Policy Tasks, the five-year plan of the Moon Administration.[53] To this end, in 2017, the Presidential Committee on the Fourth Industrial Revolution (PCFIR) was established under the direct control of the President.[54] In early 2021, the PCFIR was reorganised into a system of prime minister and private co-chairperson for gathering various opinions through close public-private cooperation.[55] As a consultative body, the principal role of the PCFIR includes deliberating and coordinating policies necessary to adapt to upcoming changes from the Fourth Industrial Revolution.[56]

The government of South Korea is also seeking to implement the Digital New Deal (a core of the Korean New Deal), a project to recover from the economic crisis caused by the covid-19 pandemic, through various means including accelerating digitalisation through the use of 5G and AI across all industries.[57] In July 2021, one year after the establishment of the Korean New Deal, the government presented a revised Korean New Deal 2.0 based on the need to urgently respond to internal and external

[52] "Master Plan Implementation Plan: Implementation Plan", Law & Policy, Personal Information Protection Commission, accessed 17 July 2021. Available at https://www.pipc.go.kr/eng/user/lgp/pbp/mlp/masterPlan.do.

[53] "Policy Information: Policy Tasks [정책정보: 국정과제]," Cheong Wa Dae [청와대], accessed 17 July 2021. https://www1.president.go.kr/government-projects.

[54] Cheong Wa Dae, "Policy Tasks".

[55] "A New Start towards A Leading Country in the Data Economy! [데이터 경제 선도국가로 향하는새로운 출발!]," (Press release, Office for Government Policy Coordination / Prime Minister's Secretariat [보도자료, 국무조정실 국무총리비서실], 17 February 2021), 2, https://www.opm.go.kr/flexer/view.do?ftype=hwp&attachNo=105133; "About PCFIR: Greetings," Presidential Committee on the Fourth Industrial Revolution, accessed 7 August 2021. https://www.4th-ir.go.kr/en/greetings.

[56] "About PCFIR: Overview," Presidential Committee on the Fourth Industrial Revolution, accessed August 7, 2021, https://www.4th-ir.go.kr/en/overview; "About PCFIR: Greetings."

[57] "What is the Digital New Deal?" Welcome to – Korean Digital New Deal, Ministry of Science and ICT, accessed 6 August 2021. available at https://digital.go.kr/front/main/eng.do#news.

changes.[58] Bolstering new digital industries and spreading the benefits from the project throughout the economy and society are the core of the plans for the Digital New Deal in the Korean New Deal 2.0.[59]

3 Shared Threats, Common Concerns and Cooperation

The EU and South Korea face high levels of cyber threats that are unique to their geopolitical and territorial context. For the EU the main concern relates to Russia, and for South Korea, North Korea and China constitute the most prominent concern. This said, they also face a number of shared challenges—and a common interest in developing more substantive cooperation mechanisms. This relates not only to countering the cyber threats posed by Russia, China and North Korea, but also in ensuring that global standards and best practices as well as values and norms are constituted that ensure a resilient, rules-based cyber ecosystem can exist to protect human rights online, as well as providing for the economic, social and political benefits that a secure and increasingly digitalised world can offer.[60]

3.1 Cybersecurity

The EU and South Korea have consistently recognised the importance of, and reiterated their commitment to, bilateral and multilateral cooperation within established formal frameworks—and in particular through the EU-South Korea Cyber Dialogue. At the November 2020 meeting, for example, the EU and South Korea "reaffirmed their commitment to an open, free, stable and secure cyberspace" and "confirmed their continued bilateral cooperation, including through annual dialogues", as well as a commitment to explore areas "for further cooperation such as … capacity building projects as well as … implementation of confidence building

[58] Jae-ho Cheon, "Government Announces Korean New Deal 2.0, An Upgrade to Properly Reflect Changes," (Press Releases, Ministry of Economy and Finance, 14 July 2021), 1. available at https://english.moef.go.kr/pc/selectTbPressCenterDtl.do?boardCd=N0001&seq=5173.

[59] Jae-ho Cheon, "Korean New Deal 2.0", 1–2.

[60] George Christou and Michael Raska, "EU–ASIA Cybersecurity Cooperation: Paths and Patterns" in *The European Union's Security Relations with Asian Partners*, Eds. Thomas Christiansen, Emil Kirchner and Tan See Seng (Springer, 2021); Ebert and Groenendaal, 5–8 and 29–30.

measures contributing to stability in cyberspace".[61] For some observers, however, a high degree of convergence in terms of threats and interests has not been translated into the need for urgent action when it comes to cooperation, even though it has created the potential to move "beyond information exchange and trust building".[62] Thus, others have noted that whilst the annual cyber dialogue "is positive in terms of reinforcing common positions in other multilateral and regional fora, more diverse and frequent interchanges should be promoted to ensure and implement practical and substantial ways to cooperate"[63] and that more joint institutionalised action could be more visible.[64]

Whilst the EU-South Korea bilateral cyber dialogue constitutes the main type of formal institutionalised cooperation there have been progress and movement towards more practical action in the technical, informal realm. At the technical level, for example, although there is EU and South Korea member cooperation in global fora such as the Forum of Incident Response and Security Teams (FIRST), there is also a commitment for enhancing direct bilateral cooperation between CERT-EU and the Korean National CERT.[65] Building on the cyber dialogue, at the informal level the EU has funded projects such as EU Cyber Direct to connect multiple stakeholders, public and private, in the discussion of cybersecurity challenges. In this context, the EU Cyber Direct project and the National Security Research Institute in South Korea organised a Track 1.5 meeting in October 2020, which included high-level officials from within the EU institutions, EU member states and South Korean ministries and agencies, as well as non-governmental experts where views were exchanged on the issues of cyber resilience, cyber conflict prevention, the geopolitics of 5G and cybercrime.[66]

[61] 6th annual European Union – Republic of Korea Cyber Dialogue, European Commission, EURAXESS, 25/11/2020. https://cdn4.euraxess.org/worldwide/south-korea/6th-annual-european-union-%E2%80%93-republic-korea-cyber-dialogue.

[62] Ebert and Groenendaal 2020, 31.

[63] Christou and Lee, 69.

[64] Christou and Raska, 210.

[65] Ebert and Groenendaal 2020, "Cyber Resilience", 31. See also: https://www.first.org/members/teams/.

[66] Ebert, Ian Wallace, Laura Groenendaal, So Jeong Kim, and Dong Hee Kim, "EU-SOUTH KOREA Cyber Consultations: Resilience and Trust in Cyberspace", EU Cyber Direct, German Marshal Fund, 16 December 2020. https://www.gmfus.org/publications/eu-rok-consultations-resilience-and-trust-cyberspace.

The EU has also sought to enhance cooperation in cybersecurity with all Asian countries, including South Korea, through its *Enhanced EU Security Cooperation in and with Asia*[67] strategy. This has resulted in a project of the same name, co-financed by the German Federal Foreign Office and the French Ministry for Europe and Foreign Affairs to be implemented jointly by Deutsche Gesellschaft für Internationale Zusammenarbeit (GIZ) GmbH and Expertise France. The project involves an initial set of six pilot countries—India, Indonesia, Japan, South Korea, Singapore and Vietnam—and seeks to explore where the EU's interests and priorities intersect with the greatest opportunities for enhanced cooperation across four security issue areas, including cybersecurity.[68]

Cooperation in the area of Digital Economy has been formalised through the Agreement for Scientific and Technological Cooperation which came into force in 2007 and established the EU-Korea Science and & Technology Joint Committee (which meets annually)—with an ICT Working Group dealing with all ICT-related issues. On cybercrime, South Korea made a decision to join the Budapest Convention in 2019, and this has been discussed formally in the political dialogue with the EU. Further cooperation, however, has been constrained by concerns about the compatibility of Korean legislation with the requirements of the Convention although work is being undertaken in South Korea to remedy the existing divergence, in particular in relation to storage of evidence, access to encrypted communication and the ability to respond to requests for evidence. On cyber conflict prevention, collaboration at bilateral level has helped to reinforce positions and secure agreement at global (UN) level on issues such as the application of international law and norms of state behaviour—within the UN Group of Governmental Experts on Developments in the Field of Information and Telecommunications (UNGGE) and Open-Ended Working Group (OEWG), for example—and there has also been cooperation on capacity building between the EU and South Korea's Global Cybersecurity Centre for Development[69] (GCCD) and Cybersecurity Alliance for Mutual Progress (CAMP).[70]

[67] Council of the European Union, Enhanced Security Cooperation in and with Asia, Council Conclusions, Brussels, 28 May 2018. https://www.consilium.europa.eu/media/35456/st09265-re01-en18.pdf.

[68] For details see: Enhanced Security Cooperation in and with Asia project, https://www.giz.de/en/worldwide/87412.html.

[69] https://cybilportal.org/actors/global-cybersecurity-center-for-development-gccd-of-korea-internet-security-agency-kisa/.

[70] https://thegfce.org/introducing-camp-cybersecurity-alliance-for-mutual-progress/.

Further cooperation on issues of cybercrime as well as privacy and data protection could potentially be impacted by the covid-19 pandemic and concerns about the use of surveillance technologies by South Korea to mitigate its effects.[71]

3.2 Data Protection and Privacy

The extraterritorial application of the GDPR has focused the adequacy issue on discussions on data protection for business transactions between the EU and South Korea. Under Article 45 of the GDPR, the European Commission may decide that data protection in a third country ensures an adequate level, and if such a decision is adopted, personal data can be transferred to the third country from the EU without any further safeguards being necessary similar to the transfer of personal data within the EU.[72] Since the Commission announced adequacy talks with South Korea in January 2017,[73] talks progressed based on the joint commitment to personal data protection as a fundamental human right.[74] Indeed the Commission, on 30 March 2021, confirmed "the high degree of convergence between the European Union and the Republic of Korea in the area of data protection" in its initial finding.[75] Enhancing PIPC's functions and independence through the aforementioned revisions of the South Korean laws, including PIPA, played a significant role in bringing about this result of the initial finding.[76] In the past, consultations for receiving adequacy status had been suspended twice because of the lack of independence of

[71] Ebert and Groenendaal, 31; Ebert et al., 2–3. See also: https://www.apc.org/en/news/era-covid-19-south-koreas-new-normal-digital-surveillance-state.

[72] "Adequacy decisions," European Commission, accessed 31 July 2021, https://ec.europa.eu/info/law/law-topic/data-protection/international-dimension-data-protection/adequacy-decisions_en.

[73] "Joint Statement by Commissioner Reynders and Yoon Jong In, Chairperson of the Personal Information Protection Commission of the Republic of Korea," European Commission, 30 March 2021. https://ec.europa.eu/commission/presscorner/detail/en/statement_21_1506.

[74] For example, see "Joint Press Statement by Mr. Lee Hyo-sung, Chairman of the Korea Communications Commission (KCC) and Mrs. Věra Jourová, Commissioner for Justice, Consumers and Gender Equality of the European Commission on the state of play of the adequacy dialogue" (1 June 2018).

[75] European Commission, "Joint Statement by Reynders and Yoon".

[76] European Commission, "Joint Statement by Reynders and Yoon".

the personal data supervisory authority.[77] After the elevation of the PIPC to an independent supervisory body, the dialogues for adequacy decision progressed more rapidly.

The initial finding excluded the processing of personal credit information by controllers such as financial institutions that are subject to the Financial Services Commission's (FSC) supervision under the Credit Information Act from the scope of granting adequacy status.[78] The ground for this exclusion was the lack of independence of the FSC as a supervisory body.[79] The fact that the FSC not only serves as a supervisory body but also has other functions such as promoting the development of financial industries was reflected in the finding.[80]

In June 2021, the European Commission sent its draft adequacy decision to the European Data Protection Board (EDPB) for review.[81] The EDPB adopted its opinion on the draft in September 2021 in which it generally affirmed the equivalence of South Korea's data protection framework with that of the EU, whilst also recommending further clarification and monitoring of certain aspects (e.g. impacts of pseudonymisation and the lack of a general right to withdraw consent) to the European

[77] Jong In Yoon (Chairperson of the PIPC) [윤종인 위원장], "Announcement of Conclusion of Discussions on Adequacy Decision between South Korea and the EU [한-EU 적정성 결정 논의 마무리 발표]," Republic of Korea Policy Briefing [대한민국 정책브리핑], 29 March 2021. https://www.korea.kr/news/policyBriefingView.do?newsId=156443808; Ebert and Groenendaal, "Cyber Resilience", 30.

[78] But the processing of personal credit information by commercial operators that are supervised by the PIPC is included in the scope of granting adequacy status in the initial finding. European Commission. "Draft Decision on the adequate protection of personal data by the Republic of Korea," 14 June 2021, 2, 4, 27–31, 63–64, and 67. Available at https://ec.europa.eu/info/files/draft-decision-adequate-protection-personal-data-republic-korea_en.

[79] Jong In Yoon (Chairperson of the PIPC), "Conclusion of Discussions on Adequacy Decision".

[80] Jong In Yoon (Chairperson of the PIPC), "Conclusion of Discussions on Adequacy Decision".

[81] European Data Protection Board. "Opinion 32/2021 regarding the European Commission Draft Implementing Decision pursuant to Regulation (EU) 2016/679 on the adequate protection of personal data in the Republic of Korea, Version 1.0" Adopted on 24 September 2021, 4. Available at https://edpb.europa.eu/our-work-tools/our-documents/opinion-art-70/opinion-322021-regarding-european-commission-draft_en.

Commission.[82] The adequacy decision, adopted on 17 December 2021,[83] allows freer transfer of personal data from the EU to South Korea. Therefore, it will facilitate business transactions between the EU and South Korea by relieving the burden on South Korean companies seeking access to the European market. It would also have the effect of complementing the Free Trade Agreement between the EU and South Korea.[84]

The GDPR requires the European Commission to continuously monitor relevant developments and conduct a periodic review at least every four years after the adoption of the adequacy decision, and if it is determined that the adequacy of the level of protection is no longer met, the decision is repealed, amended or suspended.[85] Accordingly, a first review to examine whether South Korea's legal framework and practice still ensure the adequate level of protection will be conducted within three years after the adoption.[86] Issues on maintaining adequacy status and amending the decision (e.g. the expansion of its scope to processing personal credit information which was excluded from the initial finding) may be included in future discussions in this regard.[87]

3.3 *Emerging Technologies*

On emerging technologies, cooperation remains at the level of declaration rather than practice or action. There was an agreement between the EU and South Korea in November 2013 to promote collaboration in Research and Development in ICT particularly in relation to future generation communication networks and a declaration the year after on working towards developing a global definition of 5G. Whilst this has led to a Memorandum of Understanding and the subsequent establishment of a more regular ICT Dialogue in November 2016 to discuss matters of mutual interest (5G, IoT, cloud, etc.)—it has not progressed to any form of concrete

[82] European Data Protection Board, "Opinion regarding the Draft", 1 and 5–8.

[83] European Commission, "Commission Implementing Decision", C(2021) 9316 final, 17 December 2021; at https://ec.europa.eu/info/sites/default/files/1_1_180366_dec_ade_kor_new_en.pdf.

[84] European Commission, "Joint Statement by Reynders and Yoon".

[85] Article 45(3)-(5) of the GDPR.

[86] European Commission, "Implementing Decision", 67.

[87] See European Commission, "Implementing Decision", 66–71, and 5., footnote 23; See also Jong In Yoon (Chairperson of the PIPC), "Conclusion of Discussions on Adequacy Decision".

action—although EU and South Korean companies did work together to "showcase 5G technology at the 2018 Olympic games in Pyeongchang".[88]

In June 2020 there was also commitment from EU and South Korea leaders to "strengthen cooperation in responding to the Fourth Industrial Revolution, harnessing human-centric digitalisation and promoting safe and ethical implementation of AI technologies" as well as to launch a High-Level Dialogue on the digital economy in the near future.[89] As a result, a video conference of the 1st EU-South Korea High-Level ICT Policy Dialogue was held on 18 November 2020.[90] At this conference, both sides shared their main policies on ICT, confirmed their common vision and policy directions, and discussed plans for intensifying EU-South Korea cooperation on ICT.[91] They also planned to develop the Policy Dialogue as a regular annual meeting where multiple stakeholders engage.[92]

4 Conclusions

The EU and South Korea recognise the need to address cybersecurity challenges going forward, with the covid-19 pandemic adding to the number of issues that increased digitalisation brings with it in terms of online privacy, data protection, safety and security. They are normatively like-minded in relation to the applicability of international law in cyberspace, the need to implement agreed norms for state behaviour agreed at UN level and on the need for cyber capacity building in order to facilitate the effective implementation of CBMs. They also face a number of shared threats and common interests relating to cybersecurity, data protection and privacy and emerging technologies—which has led to differentiated results in terms of action when it comes to cooperation and collaboration. There are also certain areas—such as cybercrime—where divergent legal practice constrains further effective collaboration.

Overall, whether divergent or convergent on central issues, there remains room for improvement in terms of moving beyond dialogue and declaration

[88] Ebert and Groenendaal 2020, 32.

[89] European Commission, Joint Press Release: Republic of Korea – EU Leaders' video conference meeting, 30 June 2020. https://ec.europa.eu/commission/presscorner/detail/en/IP_20_1226.

[90] "Press Releases: Start Collaboration with the EU on the ICT Policy (Nov. 18)", News&Notice, Ministry of Science and ICT, accessed 10 August 2021. https://english.msit.go.kr/eng/bbs/view.do?sCode=eng&mId=4&mPid=2&pageIndex=&bbsSeqNo=42&nttSeqNo=470&searchOpt=ALL&searchTxt=.

[91] See Ministry of Science and ICT, "Start Collaboration".

[92] Ministry of Science and ICT, "Start Collaboration".

at formal institutionalised levels to result-oriented cooperation. To this end, both formally and informally, there has been a commitment to do more in different areas and fruitful discussion of concrete ideas for further cooperation that could facilitate closing the declaration-practice gap (see Table 1).

Table 1 EU-South Korea: Key Issues and Further Cooperation

Issue area	*Further cooperation*
Cyber Resilience *Convergence on the need for a global approach to Critical Infrastructure Protection (CIP)*	Increased information sharing on risk-management and its measurement. Joint exercises and scenario planning and more joint research projects to foster further convergence towards a joint (global) approach. Further joint EU-South Korea efforts at UN level to discuss a global approach to CIP with like-minded and non-like-minded states (Ebert et al., 1).
Cyber Conflict Prevention *Convergence between the EU and SOUTH KOREA on the applicability of international law*	Increase bilateral and regional cooperation on cyber capacity building to create conditions for more effective implementation of CBMs. Closer coordination and collaboration to build alliances (and bridges) at the UN level in processes such as the OEWG (Ebert et al., 2; Ebert and Groenendaal, 33).
Cybercrime *Legally divergent*	Further exchange of information in terms of cross-border access. Further bilateral collaboration with EU member states on legislation compatible with the Budapest Convention (France, Spain, Germany). Enhanced dialogue with the Council of Europe. Enhanced cooperation in diplomatic efforts to negotiate a cybercrime treaty in the United Nations General Assembly's Third Committee. Exchange of best practice in tackling disinformation exacerbated by the covid-19 pandemic. Joint projects/investigation teams in operational matters, and multi-stakeholder events to discuss positions and how to project them (e.g. on human rights/civil liberties at UN level) (Ebert and Groenendaal, 33; Ebert et al., 3).
Data Protection/Privacy/ Adequacy *Convergent positions*	Continued, deeper bilateral discussion on outstanding/ future issues affecting adequacy.
Emerging Technologies *Shared interests and challenges*	Sharing information on certification schemes for ICT products and improving technical cooperation. Improved and deeper engagement through a multidisciplinary platform that includes political and technical experts (Ebert et al. 2020, 2).

Notwithstanding their own specific regional cyber threats, the EU and South Korea share common interests that will enhance prospects for future collaboration and cooperation in order to achieve shared normative digital goals. The broader challenge for the EU and South Korea will be to ensure not just that other like-minded states remain engaged in achieving a secure, safe, open and free Internet, but that compromises can be found with those states that have alternative visions that can potentially undermine such a vision if consensual norms for state and non-state actor behaviour in cyberspace are not implemented beyond discursive outcome.

References

"A New Start towards a Leading Country in the Data Economy! [데이터 경제 선도국가로 향하는 새로운 출발!]," Press release. Office for Government Policy Coordination / Prime Minister's Secretariat [보도자료, 국무조정실 국무총리비서실], 17 February 2021. https://www.opm.go.kr/flexer/view.do?ftype=hwp&attachNo=105133 (accessed 7 August 2021).

2020 Annual Report on Personal Information Protection. Personal Information Protection Commission, 2020. https://www.pipc.go.kr/np/ebook/2020Y_report/index.html (accessed 17 July 2021).

Carrapico, Helena and Farrand, Benjamin. "Discursive continuity and change in the time of Covid-19: the case of EU cybersecurity policy", *Journal of European Integration*, 42, Issue 8 (2020), 1111–1126.

Cheon, Jae-ho. "Government Announces Korean New Deal 2.0, An Upgrade to Properly Reflect Changes," Press Releases, Ministry of Economy and Finance, 14 July 2021. https://english.moef.go.kr/pc/selectTbPressCenterDtl.do?boardCd=N0001&seq=5173 (accessed 7 August 2021).

Cheong Wa Dae [청와대]. "Policy Information: Policy Tasks [정책정보: 국정과제]." https://www1.president.go.kr/government-projects (accessed 27 July 2021).

Christou, George. "The Collective Securitisation of EU Cybersecurity", *West European Politics*, 42, Issue 2 (2019), 278–301.

Christou, George *Cybersecurity in the European Union: Resilience and Adaptability in Governance Policy* (Houndmills, Basingstoke: Palgrave Macmillan, 2016).

Christou, George and Ji Soo Lee. "EU–ROK cooperation on cyber-security and data protection" in *EU-Korea Security Relations*, ed. Nicola Casarini (London: Routledge, 2021), 55–77.

Christou, George and Michael Raska. "EU–ASIA Cybersecurity Cooperation: Paths and Patterns" in *The European Union's Security Relations with Asian Partners*, eds. Thomas Christiansen, Emil Kirchner and Tan See Seng (Springer, 2021).

Council of the European Union. "Council Conclusions on Shaping Europe's Digital Future", 2020a (No. OJ 2020C: 202).

Council of the European Union. "Council Conclusions on Covid-19", 2020b (No. 2020/C 57/4).

Council of the European Union. Enhanced Security Cooperation in and with Asia, Council Conclusions, Brussels, 28 May 2018. https://www.consilium.europa.eu/media/35456/st09265-re01-en18.pdf (accessed 22 July 2021).

Council of the European Union. "Shared Vision, Common Action: A Stronger Europe. A Global Strategy for the EU's Foreign and Security Policy", 2016. https://eeas.europa.eu/archives/docs/top_stories/pdf/eugs_review_web.pdf (accessed 25 July 2021).

Dekker, Brigitte and Okano-Heijmans, Maaike. "The Implications of China's Techno-Nationalism for the European Union and the Republic of Korea, Europe and the World Articles, Clingendael Magazine, 27 October 2020. https://www.clingendael.org/publication/eu-republic-korea-digital-connectivity-unitedwe-must-stand (accessed 25 July 2021).

Ebert, Hannes and Groenendaal, Laura. "Cyber Resilience and Diplomacy in the Republic of Korea: Prospects for EU Cooperation, Digital Dialogue", EU Cyber Direct, (2020). https://eucd.s3.eu-central-1.amazonaws.com/eucd/assets/iXvfE20z/digital-dialogue-rok.pdf (accessed 25 July 2021).

Ebert, Hannes, Wallace, Ian, Groenendaal, Laura, Kim, So Jeong and Kim, Dong Hee. "EU-ROK Cyber Consultations: Resilience and Trust in Cyberspace", EU Cyber Direct, German Marshal Fund, 16 December 2020. https://www.gmfus.org/publications/eu-ROK-consultations-resilience-and-trust-cyberspace (accessed 25 July 2021).

Enhanced Security Cooperation in and with Asia project. https://www.giz.de/en/worldwide/87412.html (accessed 25 July 2021).

ENISA. Threat Landscape Report, 2020. https://www.enisa.europa.eu/topics/threat-risk-management/threats-and-trends?tab=details (accessed 27 July 2021).

European Commission. "Adequacy decisions", 2021a. https://ec.europa.eu/info/law/law-topic/data-protection/international-dimension-data-protection/adequacy-decisions_en (accessed 31 July 2021).

European Commission. "Commission Implementing Decision", C(2021) 9316 final, 17 December 2021b. https://ec.europa.eu/info/sites/default/files/1_1_180366_dec_ade_kor_new_en.pdf.

European Commission. 6th annual European Union – Republic of Korea Cyber Dialogue, EURAXESS, 25 November 2020a. https://cdn4.euraxess.org/worldwide/south-korea/6th-annual-european-union-%E2%80%93-republic-korea-cyber-dialogue (accessed 27 July 2021).

European Commission and High Representative of the Union for Foreign Affairs and Security Policy (2020). Joint Communication to the European Parliament

and the Council: The EU's Cybersecurity Strategy for the Digital Decade, Brussels, 16.12.2020 JOIN (2020) 16 final, 1–28.
European Commission. "Shaping Europe's Digital Future"; European Commission. "A European Strategy for Data", 2020b (No. COM (2020) 66).
European Commission. "A New Industrial Strategy for Europe", 2020c (No. COM (2020) 102 F).
European Commission. Communication from the Commission to the European Parliament, The European Council, The Council, The European Economic and Social Committee and the Committee of the Regions on the "EU Security Union Strategy", 2020d, Brussels, 24.7.2020, COM (2020) 605 final.
European Commission. "Draft Decision on the adequate protection of personal data by the Republic of Korea." 14 June 2021c. https://ec.europa.eu/info/files/draft-decision-adequate-protection-personal-data-republic-korea_en (accessed 28 July 2021).
European Commission. "Joint Statement by Commissioner Reynders and Yoon Jong In, Chairperson of the Personal Information Protection Commission of the Republic of Korea." 30 March 2021d. https://ec.europa.eu/commission/presscorner/detail/en/statement_21_1506 (accessed 28 July 2021).
European Commission. "Joint press release: Republic of Korea -EU Leaders' video conference meeting", June 30, 2020e. https://ec.europa.eu/commission/presscorner/detail/en/IP_20_1226 (accessed 27 July 2021).
European Commission. Proposal for a Regulation of the European Parliament and of the Council concerning the respect for private life and the protection of personal data in electronic communications and repealing Directive 2002/58/EC (Regulation on Privacy and Electronic Communications) COM/2017/010 final – 2017/03 (COD).
European Commission. "European agenda on security", 28 April, COM (2015) 185 final. https://ec.europa.eu/home-affairs/sites/homeaffairs/files/e-library/documents/basicdocuments/docs/eu_agenda_on_security_en.pdf (accessed 29 July 2021).
European Commission and High Representative of the European Union for Foreign Affairs and Security Policy/Vice-President of the Commission. "Cybersecurity Strategy of the EU: An Open, Safe and Secure Cyberspace", Brussels, 7 February 2013.
European Commission. "Proposal for a regulation of the European Parliament and of the Council on the protection of individuals with regard to the processing of personal data and on the free movement of such data (General Data Protection Regulation)", 2012, COM (2012) 11 final, Brussels, 25.1.2012, 2.
European Commission. "European principles and guidelines for Internet resilience and stability", March 2011. http://ccpic.mai.gov.ro/docs/guidelines_internet_fin.pdf (accessed 29 July 2021).
European Commission. Communication from the Commission to the European Parliament, the Council, the European Economic and Social Committee and

the Committee of the Regions, "A Digital Agenda for Europe", 2010a (COM (2010) 245 Final/2), Brussels.

European Commission. "The EU internal Security Strategy in Action: Five Steps Towards a More Secure Europe", 2010b (No. COM (2010) 673 Final), Brussels.

European Commission and European External Action Service. Joint Communication to the European Parliament and the Council, Joint Framework on Countering Hybrid Threats: A European Union response. 6 April, JOIN (2016) 18 final.

European Data Protection Board. "Opinion 32/2021 regarding the European Commission Draft Implementing Decision pursuant to Regulation (EU) 2016/679 on the adequate protection of personal data in the Republic of Korea, Version 1.0" Adopted on 24 September 2021. https://edpb.europa.eu/our-work-tools/our-documents/opinion-art-70/opinion-322021-regarding-european-commission-draft_en (accessed 24 October 2021).

EUROPOL. Internet Organised Crime Threat Assessment, 2020. https://www.europol.europa.eu/activities-services/main-reports/internet-organised-crime-threat-assessment-iocta-2020 (accessed 27 July 2021).

Fahey, Elaine. "The EU's cybercrime and cyber-security rulemaking: mapping the internal and external dimensions of EU security", European Journal of Risk Regulation, 5, Issue 1 (2014), 46–60.

Gaumond, Eve. "Artificial intelligence Act: What is the EU Approach for AI?", LAWFARE.

"Joint Press Statement by Mr. Lee Hyo-sung, Chairman of the Korea Communications Commission (KCC) and Mrs. Věra Jourová, Commissioner for Justice, Consumers and Gender Equality of the European Commission on the state of play of the adequacy dialogue." 1 June 2018.

Master Plan [개인정보 보호 기본계획] (2021–2023). Personal Information Protection Commission, November 2020. https://www.pipc.go.kr/np/default/page.do?mCode=D050010010#LINK (accessed 27 July 2021).

Ministry of Science and ICT. "Press Releases: Start Collaboration with the EU on the ICT Policy (Nov. 18)", News & Notice. https://english.msit.go.kr/eng/bbs/view.do?sCode=eng&mId=4&mPid=2&pageIndex=&bbsSeqNo=42&nttSeqNo=470&searchOpt=ALL&searchTxt= (accessed 10 August 2021a).

Ministry of Science and ICT. "What is the Digital New Deal?," Welcome to – Korean Digital New Deal. https://digital.go.kr/front/main/eng.do#news (accessed 6 August 2021b).

2021 National Cybersecurity White Paper (Korean version). National Intelligence Service (NIS) et al., 2021.

National Security Strategy. National Security Office, 2018.

National Cybersecurity Strategy. National Security Office, 2019.

Personal Information Protection Commission. "Master Plan· Implementation Plan: Implementation Plan." Law & Policy. https://www.pipc.go.kr/eng/user/lgp/pbp/mlp/masterPlan.do (accessed 17 July 2021a).

Personal Information Protection Commission. "Master Plan· Implementation Plan: Master Plan." Law & Policy. https://www.pipc.go.kr/eng/user/lgp/pbp/mlp/masterPlan.do (accessed 17 July 2021b).

Presidential Committee on the Fourth Industrial Revolution. "About PCFIR: Greetings." https://www.4th-ir.go.kr/en/greetings (accessed 7 August 2021a).

Presidential Committee on the Fourth Industrial Revolution. "About PCFIR: Overview." https://www.4th-ir.go.kr/en/overview (accessed 17 July 2021b).

Reiterer, Michael. "The 10th anniversary of the EU-Korea Strategic partnership", VUB-KF Chair, Policy Brief, Issue 2020/09, July 2020. https://www.korea-chair.eu/wp-content/uploads/2020/07/KFVUB_Policy-Brief-2020-09.pdf (accessed 31 July 2021).

Republic of Korea[관계부처 합동]. *National Cybersecurity Basic Plan[국가 사이버안보 기본계획]* 2019. https://policy.nl.go.kr/search/searchDetail.do?rec_key=SH2_PLC20190240687 (accessed 17 July 2021).

Vestager, Margrethe. Executive Vice-President for a Europe fit for the Digital Age, "Europe fit for the Digital Age: Commission proposes new rules and actions for excellence and trust in Artificial Intelligence", 2021. https://digital-strategy.ec.europa.eu/en/news/europe-fit-digital-age-commission-proposes-new-rules-and-actions-excellence-and-trust-artificial (accessed 27 July 2021a).

Vestager, Margrethe. Executive Vice-President for a Europe fit for the Digital Age, "Answers to the European Parliament questionnaire to the Commissioner-designate", 4–6. https://ec.europa.eu/commission/commissioners/sites/default/files/commissioner_ep_hearings/answers-ep-questionnaire-vestager.pdf (accessed 27 July 2021b).

Yoon, Jong In. (Chairperson of the PIPC). "Announcement of Conclusion of Discussions on Adequacy Decision between South Korea and the EU." Republic of Korea Policy Briefing [윤종인 위원장. "한-EU 적정성 결정 논의 마무리 발표." 대한민국 정책브리핑], Mar. 29, 2021. https://www.korea.kr/news/policyBriefingView.do?newsId=156443808 (accessed 17 July 2021).

Cyber Offence Dominance, Regional Dynamics, and Middle Power–led International Cooperation

Mason Richey

1 Introduction

Cyberspace is not like other security domains. It is novel, inherently transnational, and dematerialized; its weapons (e.g., computer code) are cheap and rapidly adaptable; many cyber actors are unidentifiable; many attacks are deniable and/or difficult to attribute; cyber governance via international law is weak. Institutionalization of international cyber norms and

M. Richey (✉)
Hankuk University of Foreign Studies, Seoul, South Korea
e-mail: mrichey@hufs.ac.kr

© The Author(s), under exclusive license to Springer Nature Switzerland AG 2022
G. Boulet et al. (eds.), *Cybersecurity Policy in the EU and South Korea from Consultation to Action*, New Security Challenges,
https://doi.org/10.1007/978-3-031-08384-6_4

rules is inchoate and patchwork.[1] Few of these efforts at institutionalization have won even the support of all the states involved in promulgating the international governance structures (laws, guidelines, best-practices). States without a seat at the negotiation table have been even less inclined to follow these international cybersecurity norms and regulations.

In short, there is no broad, coherent multilateral regime regulating cybersecurity. A fortiori, as concerns national and military security, there are no truly multilateral conventions governing cyber conflict and cyber operations. This lacuna persists despite the growth of the cyber domain (including artificial intelligence [AI]) as an important area of contestation and conflict vector. Indeed cyber is already commonplace for grey-zone strategies, most obviously in terms of cyber espionage and political subversion.[2] Moreover cyber is increasingly integrated into other security domains. Cyber technology and AI now interface with air and space domains via unmanned aerial vehicles (UAVs), satellite communications and sensoring, and nuclear weapons and missile defence technologies; with the land domain via robotics and logistics and supply-chain

[1] For example, UN 1st Committee on ITU and 2021 76th session on Agenda item 95 (Developments in the field of information and telecommunications in the context of international security); 2001 Convention on Cybercrime; EU Directive on NIS Security; NATO Tallinn Manuals and others mentioned in this volume.

[2] Gary Corn (2019). "National Cyber Security: Navigating Gray-Zone Challenges in and through Cyberspace." In *Complex Battlespaces: The Law of Armed Conflict and the Dynamics of Modern Warfare* (eds. Winston Williams and Christopher Ford). (Oxford: Oxford University Press); Mark Emem (2019). "North Korea's $700 million Crypto Theft Chest Could Fund Nukes: Report." *CNN* (April 15). (https://www.ccn.com/north-koreas-700-million-crypto-theft-chest-could-fund-nukes-report/); Hannah Murphy (2019). "Cyber security companies race to combat 'deepfake' technology." *Financial Times* (August 16). (https://www.ft.com/content/63cd4010-bfce-11e9-b350-db00d509634e).

optimization; and with the maritime domain via naval electronic warfare and unmanned systems.[3] Militaries everywhere are recruiting cyberwarriors.[4]

The lack of international laws or regimes covering cybersecurity—broadly construed from cybercrime to cyberwar—is problematic both because lack of norms and norm enforcement incentivizes predatory behaviour, and because lack of consensus on definitions of cyber attacks, understanding of red lines, and agreement on acceptable cyber defence behaviour and (if necessary) proportionate retaliation increases risks of misperception, misunderstanding, cybersecurity dilemmas, and inadvertent conflict escalation. Moreover these risks are prevalent at a moment in which offensive cyber operations are superior to defensive strategies, in significant part because deterrence has limited effectiveness in cyberspace. The lack of a coherent international governance architecture for cybersecurity is a permissive condition for aggressive cyber actors to capitalize on the offensive dominance of cyber technology. This problematic confluence of an international cybersecurity regulatory vacuum and cyber offence dominance is exacerbated by a multipolar international system, which tends towards greater instability than bipolar systems.[5]

It is in this context of the cyber domain that this chapter analyses strategic competition and conflict among selected states. More specifically,

[3] Sandra Erwin (2018). "Defense inspector general finds key Air Force space programs vulnerable to cyber attacks, sabotage." *Space News* (August 16). (https://spacenews.com/defense-inspector-general-finds-key-air-force-space-programs-vulnerable-to-cyber-attacks-sabotage/); Patricia Lewis and Beyza Unal (2019). "The Destabilizing Danger of Cyberattacks on Missile Systems." Chatham House—Expert Comment (April 2). (https://www.chathamhouse.org/expert/comment/destabilizing-danger-cyberattacks-missile-systems); Sue Halpern (2019). "How Cyber Weapons Are Changing the Landscape of Modern Warfare." *The New Yorker* (July 18). (https://www.newyorker.com/tech/annals-of-technology/how-cyber-weapons-are-changing-the-landscape-of-modern-warfare); Lora Saalman (ed.) (2019). *The Impact of Artificial Intelligence on Strategic Stability and Nuclear Risk* (Vol. II—East Asian Perspectives). SIPRI Report (October).

[4] Nina Kollars and Emma Moore (2019). "Every Marine A Blue-Haired Quasi-Rifleperson?" *War on the Rocks* (August 21). (https://warontherocks.com/2019/08/every-marine-a-blue-haired-quasi-rifleperson/).

[5] Kenneth Waltz (1979). *Theory of International Politics*. (Reading, MA: Addison-Wesley).

this chapter applies offence-defence[6] theory to the cyber domain in order to predict states' likely cyberspace behaviour. Although cyber capabilities are digital (i.e., dematerialized and thus not necessarily territorially specific), this chapter examines the cyber behaviour of states primarily in the multipolar East Asia region, which is experiencing intense strategic competition among its main actors (the US, China, South Korea, North Korea, Japan, Russia). By applying Christensen's and Snyder's (1990) offence-defence theory to an unstable multipolar East Asia international system with numerous alliance relationships, one sees how the region's states are likely to "chaingang" in the event of cyberconflict (i.e., over-balancing in which one alliance partner provides near-unconditional support for and participation in another alliance partner's conflict.).[7] Given that the lack of international cybersecurity laws and regimes is a permissive condition for cyber aggression, the chapter seeks to determine how potential cybersecurity dilemmas and cyberconflict can be mitigated by developing better international cooperation—including via international institutions—devoted to cyber issues in general and cybersecurity threat reduction in particular.

This chapter proceeds as follows. Section 2 surveys the cyber threat landscape. Section 3 empirically discusses the offence dominant nature of the cyber domain, arguing that cyber's offensive dominance stems from lack of geographical obstacles to offensive cyber operations, technological and cost advantages to aggressors, attribution problems and weak cyber deterrence. The section especially focuses on how attribution problems and technological advances in cyber weapons vex cyber deterrence and create crisis instability. Section 4 applies to the cyber domain Christensen's and Snyder's (1990) theoretical analysis of state behaviour under conditions of multipolarity and perception of offensive dominance. The case examined under these parameters is East Asia, a region with rivalrous great powers, a revisionist North Korea, and three other developed, technologically sophisticated states (Russia, South Korea, and Japan)—all of which have formidable cyber capabilities, and some of which are in alliances with

[6] Charles Glaser and Chaim Kaufmann (1998). "What is the Offense-Defense Balance and Can We Measure It?" *International Security* 22(4): 44–82; Keir Lieber (2011). "Mission Impossible: Measuring the Offense-Defense Balance with Net Military Balance." *Security Studies* 20: 451–459; Rebecca Slayton (2017). "What Is the Cyber Offense-Defense Balance? Conceptions, Causes, and Assessment." *International Security* 41(3): 72–109.

[7] Thomas Christensen and Jack Snyder (1990). "Chained Gangs and Passed Bucks: Predicting Alliance Patterns in Multipolarity." *International Organization* 44(2): 137–168.

other regional states. Ultimately this section argues that one should expect states to "chaingang" in such a situation, which entails risk for widening and escalating cyberconflict. Section 5 concludes with reflections on how international institutions (at bilateral, regional, and global levels) can improve cybersecurity, notably through promotion of a regime of information exchange and best-practices establishing expectations of (and limits on) tolerable cyber behaviour, including for more proactive cyber defence. The section also explains the role that South Korea and the EU—both middle powers and strategic partners with an ongoing high-level Cyber Dialogue—can play in this regard.

2 Cyber Threat Overview

In 2021 alone, in total there were more than 300 million ransomware attacks, 82 million cryptojacking attacks, and 57 million "Internet-of-Things" (IoT) malware attacks.[8] The frequency of ***major*** system breaches—"cyber attacks against government agencies, defence and high-tech companies, and economic crimes worth more than 1mnUSD"—indicates high levels of ambitiously aggressive malicious cyber operations.[9] CSIS reports that during the 2006–2019 period there were 444 such cyber attacks.[10] Of those 444 cyber attacks, 340 involved states—either as aggressor or as target—in East Asia: Russia (152 instances), the US (101 instances), China (52 instances), North Korea (41 instances), South Korea (23 instances), and Japan (13 instances). CSIS's partial report for 2020–2021 shows these trends accelerating.[11] The true number of attacks is higher than reported figures, as targets have incentives to remain silent.

Beyond the sheer number of cybersecurity breaches, the breadth of attack types is also significant. One can think of cyber attacks under at least four broad categories of goals (which can be hybrid): (1) illicit currency acquisition, (2) sabotage, (3) strategic and tactical system intrusions

[8] SonicWall (2021). Cyber Threat Report. (https://www.sonicwall.com/2021-cyber-threat-report/).

[9] CSIS (2019). Significant Cyber Incidents. CSIS Report (September 4). (https://www.csis.org/programs/technology-policy-program/significant-cyber-incidents).

[10] Ibid.

[11] CSIS (2021). Significant Cyber Incidents. CSIS Report (April 26). (https://csis-website-prod.s3.amazonaws.com/s3fs-public/210326_Significant_Cyber_Events.pdf?ZKJIdGVXdQd2vXW.gFEcFQs2Ay7cDiqt).

(including cyber intelligence operations) and (4) political subversion and influence campaigns.

(1) Earning money through hacking is a common objective for malicious cyber actors, including especially non-state actors, although some states are major perpetrators also. Banking trojans and other traditional online banking malware are still workhorses of digital currency theft (see Kyu-dok **Hong** and Seong-jong **Song** in this volume), with cryptotheft—the illegal access into and theft of cryptocurrency from blockchain wallets—increasingly a primary source of illegal financial gain. Cryptojacking—the use of malware to hijack computers for cryptocurrency mining—is a growing frontier for illicit currency acquisition. What tends to get the most publicity, however, is ransomware: malware (trojan horses such as cryptoworms or cryptoviruses) injected into a system that is (a) subsequently encrypted to prevent file access by the system's owners, and (b) script written to self-replicate and infect other systems in a network. The cyber attackers then extort ransom (usually paid in cryptocurrency) from victims in exchange for the decryption key. Some of the best-known cyber attacks are of this variety, notably the WannaCry and Petya/NonPetya attacks in 2017 (see also Joohui **Park** in this volume) and the 2021 Colonial Pipeline hack, in which Darkside malware forced Colonial Pipeline to shut down the largest oil pipeline on the US East Coast (carrying 45% of its fuel supply).[12] Arguably the most infamous instance of ransomware, however, was not primarily related to financial gain: the 2014 Sony Pictures hack, perpetrated by North Korea's Lazarus Group (under the name Guardians of Peace), an APT (Advanced Persistent Threat) also responsible for WannaCry. The Sony breach—which brought down the company's computer systems, stole personal employee information, and leaked embarrassing internal company communications—was (partially successfully) aimed at coercing Sony Pictures into withdrawing the theatre release of a film, *The*

[12] Ken Dilanian and Kelly O'Donnell (2021). "Russian criminal group suspected in Colonial pipeline ransomware attack." *NBC News* (May 10). (https://www.nbcnews.com/politics/national-security/russian-criminal-group-may-be-responsible-colonial-pipeline-ransomware-attack-n1266793).

Interview, that mocked North Korea and its new leader Kim Jong Un.

(2) Cybersabotage exploits include cyber-kinetic attacks, cyber "left-of-launch/boom" attacks, and cyber-infrastructure attacks (including cyber-based infiltration and sabotage of technology and/or critical systems). There is much about these dark arts that is non-disclosed, but proof-of-concept has been achieved for all these techniques.[13] More generally, crossing the cyber-physical divide aims at more than just data or computer hardware destruction, instead using algorithms to control physical systems remotely and operate them beyond their parameters so they are damaged or destroyed. The most famous example was the US-Israel designed cyber-kinetic attack Stuxnet, a worm that targeted Iranian SCADA systems and succeeded in damaging part of that country's nuclear programme by manipulating valves on uranium enrichment cascades.[14] Industrial control systems for manufacturing, energy production/distribution, water and transit, and so on are also at risk, with the IoT a growing vulnerability.

Cyber "left-of-launch"/"left-of-boom" refers to an attacker's ability to access an adversary's military platforms (especially missile/rocket or explosive) with code that interferes with or destroys them before they launch (if a missile/rocket) or detonate on target (if a bomb/warhead). This cyber operation is, strictly speaking, a part of counter-force cyberwarfare, but it is likely that securing access to such systems requires peacetime code "pre-positioning" (e.g., backdoors) or infiltration of a software supply chain used in the platform's production. In any event, such a strategy is uncertain and risky, although reportedly has been successfully used by the US to thwart North Korean ballistic missile tests.[15]

[13] For systems relying on software—especially in the security/defence sector, as well as critical systems—one strategy is to target software elements that go into a facility's operating system, without the software's users being aware that the software supply chain was compromised.

[14] Kim Zetter (2015). *Countdown to Zero Day: Stuxnet and the Launch of the World's First Digital Weapon*. (New York: Crown Publishing).

[15] Ankit Panda (2018). "US 'Left of Launch' Cyber Effort Might Increase Korean Peninsula Nuclear Dangers." *The Diplomat* (October, 22). (https://thediplomat.com/2018/10/us-left-of-launch-cyber-efforts-might-increase-korean-peninsula-nuclear-dangers/).

Cyber-infrastructure attacks aim to access a society's infrastructural systems (transit, energy, water, medical, etc.) to gain control over them and, at an appointed time, degrade (not destroy) their function. For instance, in 2015 Ukrainian utility companies were hit by (likely Russian-originated) BlackEnergy malware that infected SCADA systems and left hundreds of thousands of customers without power for six hours.[16] In October 2020, during border tensions with India, a Chinese-sponsored hacking group (Red Echo) reportedly accomplished a similar breach into Mumbai's power grid.[17] A variation on cyber-infrastructure attacks involves combining target system intrusion and encryption with ransom demands (a "double extortion" attack, in which payment by the victim is required both for the decryption key and a promise not to release sensitive files or degrade infrastructure).

(3) Strategic and tactical system intrusions comprise various hacks: mapping and understanding a target's computer systems, personal information theft and exfiltration, cyber intelligence operations. These attacks can provide an aggressor with information about a target's information systems, personnel, sensitive activities, proprietary knowledge, trade/business and government secrets, and plans and operations. Operation Cloud Hopper—attributed to a China-backed hacker group—is a sophisticated, large-scale example (see also Joohui **Park** in this volume), as is the Pegasus spyware from NSO (an Israeli company).[18] Information acquired about personnel can be leaked to embarrass or coerce/blackmail members of government or major businesses. Particularly worrisome is that strategic system intrusions may lie dormant, only in case of conflict to be used for a specific operation. This is, for instance, one of the

[16] Robert Lipovsky and Anton Cherapanov (2016). "BlackEnergy trojan strikes again: Attacks Ukrainian electric power industry." WeLiveSecurity (January 4). (https://www.welivesecurity.com/2016/01/04/blackenergy-trojan-strikes-again-attacks-ukrainian-electric-power-industry/).

[17] Tobias Burgers and David Farber (2021). "China's Dangerous Step Toward Cyber Conflict." *The Diplomat* (March 12). (https://thediplomat.com/2021/03/chinas-dangerous-step-toward-cyber-war/).

[18] Jack Stubbs et al. (2019). "Inside the West's failed fight against China's 'Cloud Hopper' hackers." *Reuters* (June 26). (https://www.reuters.com/investigates/special-report/china-cyber-cloudhopper/); David Pegg and Sam Cutler (2021). "What is Pegasus spyware and how does it hack phones?" *The Guardian* (July 18). (https://www.theguardian.com/news/2021/jul/18/what-is-pegasus-spyware-and-how-does-it-hack-phones).

concerns about the massive December 2020 APT29 (Russian "Cozy Bear") zero-day exploits against SolarWinds, Fireye, and US federal government ministries.[19]

(4) Cyber political subversion and disinformation/influence campaigns are digital operations undertaken by (principally) states and non-state actors to undermine the political stability and coherence of a target state (usually an adversary). These subversive actions and influence operations can be legal (via, e.g., social network manipulation, production/distribution of "deep fakes"), even if perfidious, but also may involve illegal incursions into computer systems to pilfer and release sensitive documents, steal personal information of political actors, or hack social media accounts to publicize falsified statements or images.

Russia—through APT28 ("Fancy Bear") and APT29 ("Cozy Bear")—is typically cited as the most active threat in this regard, with credible reports of not only legal cyber subversion in the US and Europe, but also a history of wide-ranging hacking operations to gain access to sensitive information of political campaigns and figures, as well as government offices and ministries, in the US and throughout Europe. China also allegedly has a record of these types of cyber political subversion actions, with ostensible attacks against Taiwan and India.[20] For its part, the US also has these capabilities—indeed former president Trump's National Security Memorandum 13 promulgated possible US offensive cyber operations, including cyber-enabled political subversion and influence campaigns.[21]

Cyber attacks are not limited by geography, but Eurasia—housing both aggressors and targets—is a hotspot for cybersecurity breaches, as noted in the aforementioned CSIS report. This makes sense, as most of the world's most offensively capable cyber actors—China, Russia, the UK, North Korea, and Israel, as well as, to a lesser degree, France, Spain, Germany, and the Netherlands—are resident in these two regions, while the world's

[19] Lily Newman (2020). "No one Knows How Deep Russia's Hacking Rampage Goes." *Wired* (December 14). (https://www.wired.com/story/russia-solarwinds-supply-chain-hack-commerce-treasury/).

[20] Burgers and Farber (2021).

[21] Federation of American Scientists. NSPMs—Donald J. Trump Administration. (https://fas.org/irp/offdocs/nspm/index.html).

most comprehensive (offensive and defensive) cyber power, the US, has a significant Eurasian presence.[22]

Capability alone matters, but even more important is that Eurasia is the primary venue for competition involving major powers such as the US, China, and Russia, as well as revisionist North Korea—each of whom uses cyber tools to advance their perceived interests. As the old adage goes, "when the elephants fight, the grass gets trampled." It is thus important that middle powers coordinate and cooperate, both bilaterally and multilaterally, to attenuate the ravages of the elephants. Two such middle powers are South Korea and the EU (and its member states), who have been consistent targets of cyber attacks—covering nearly all of the aforementioned threat areas—by Russia, China, and North Korea. One response has been to cooperate with each other, sharing best-practices, threat and risk perceptions, and strategic intentions both in multilateral contexts and bilaterally via a South Korea-EU Cyber Dialogue (more below and throughout this volume).[23]

3 The Cyber Offence-Defence Balance

The success of major cybersecurity incursions shows cyber's offensive superiority. Indeed cyber is the only conflict domain with near consensus perceiving offence as superior to defence, both generally and in East Asia specifically.[24] This point about the operational aspect of the cyber offence-defence balance is related to a different interpretation of the same data: deterrence also functions poorly. The reasons for this state of affairs are several.

First, unlike in the physical domain in which geography is an obstacle to offence, geography is not critical to the cyber domain. For the physical environment, accomplishing an offensive goal is tendentially harder when applying force across distance. This is due to greater supply-line fragility, increased communication difficulties, decreasing accuracy of offensive firepower and so on. For cyber attacks, however, this issue is greatly attenuated, albeit not eliminated. Cyber aggressors still face the problem of

[22] Julia Voo et al. (2020). ***National Cyber Power Index***. Harvard Belfer Center (https://www.belfercenter.org/sites/default/files/2020-09/NCPI_2020.pdf).

[23] Hannes Ebert and Laura Groenendaal (2020). "Cyber Resilience and Diplomacy in the Republic of Korea: Prospects for EU Cooperation." EU Cyberdirect Digital Dialogue. (https://eucyberdirect.eu/wp-content/uploads/2020/08/digital-dialogue-rok.pdf).

[24] Slayton (2017).

accessing a foreign system, just as militaries in physical domains have to access foreign territory in which the defender has an advantage in understanding the terrain. Temporal features of cyber attacks—including patterns and micro-time lags due to distance between attack source and target system—can alert cybersecurity experts (and AI-enabled cyberdefence software) to breaches, and are thus another minor way that geography plays a role in the cyber domain.[25] Overall, however, the digital nature of cyberspace makes for weak geographic constraints on offensive cyber exploits.

Beyond dematerialized geography, three other strategic factors contribute to superiority of cyber offence over defence: the offence-defence balance for cyber technology and cost, attribution problems, and structural weaknesses in cyber deterrence. To begin with cyber technology and cost, technological advancements in the cyber domain favour attack because of rapid development/acquisition times for offensive cyber tools.[26] For example, these can be bought "off the shelf," and custom computer code can be created quickly. The latter capability is important in the cat-and-mouse game between offensive exploiters and defenders playing technological catch-up. There are several reasons for this imbalance: cyber weapons require neither large industrialization to achieve scale effects, nor complex supply-chains of physical products subject to delays attendant to the analogue world. Offensive cyber tools can also be built and/or used by small teams (relative to traditional weapons), which are moreover the main input into the weapons' development. Taken together, these characteristics make cyber weapon development rapid, flexible, and comparatively inexpensive. Finally, on the defensive side of the cost ledger, defending against cyber attacks is difficult both because of the challenges presented by rapid, flexible offensive cyber capabilities, and because of the sheer amount of systems vulnerability, a consequence of society's reliance on digitally networked (especially Internet) technology, which will become

[25] Meir Kalech (2019). "Cyber-attack detection in SCADA systems using temporal pattern recognition techniques." *Computers and Security* 84 (July): 225–238.

[26] Christopher Bartos (2016). "Cyber Weapons Are Not Created Equal." *Proceedings of the US Naval Institute* (Vol. 142, June). (https://www.usni.org/magazines/proceedings/2016/june/cyber-weapons-are-not-created-equal).

more ubiquitous with IoT and 5G/6G.[27] The corollary to this point is that good cyber defence is expensive because it would require wholesale changes to the organization of modern society.

One of the most notorious difficulties in cybersecurity is the attribution problem.[28] This refers to the difficulty of tracking, identifying, and holding accountable the perpetrator(s) of a cyber incident or hacking exploit (see also Joohui **Park** in this volume). The attribution problem is most frequently associated with retribution dilemmas faced by targeted entities post hoc. There are, however, ex ante dimensions factoring into how the attribution problem supports the dominance of offence in cyberspace. In principle, this is a form of intelligence failure, as it is hard to locate and observe cyber attackers due to their easy concealment (small footprint, little materiel, potentially global location) and the fact that the Internet's architecture provides ways for cyber aggressors to dissimulate their activity.[29] Moreover, unlike the physical world—in which military build-up, distinct signatures to weapons development and testing, and large budget allocation are noticeable—malicious cyber actors can develop and test weapons inconspicuously.

Clearly the ex ante aspect of cyber attribution problems favours offence, as it is hard to observe, and potentially pre-empt or interdict, a cyber attack in nuce (although see Anna-Maria **Osula** in this volume on cyber resilience via disruptive, potentially pre-emptive cyber defence tech). The post hoc dimension of attribution problems—which gets more attention—is beset with uncertainty and thus also devilish. After a cyber attack, attribution requires a mixture of techniques, tools, and methods. Frequently this involves deciphering attacker intent/motive, examining code/script patterns, recognizing revealing mistakes,

[27] Ibid. Slayton (2017) argues that cost is not a critical factor in the cyber offence-defence balance, as the relative costs of both aspects of the domain are very low. She also shows that offensive cost itself is not monotonic, but depends on the objective. A standard hacking operation favours offence on cost grounds, but a complex operation (e.g., Stuxnet) is very expensive.

[28] Slayton (2017); John Davis II et al (2017). *Stateless Attribution: Toward International Accountability in Cyberspace.* (Santa Monica, CA: RAND Corporation). (https://www.rand.org/pubs/research_reports/RR2081.html); Office of the United States Director of National Intelligence (2018). A Guide to Cyber Attribution. (https://www.dni.gov/files/CTIIC/documents/ODNI_A_Guide_to_Cyber_Attribution.pdf).

[29] Joshua Rovner (2019). "Cyber War as an Intelligence Contest." *War on the Rocks* (September 16). (https://warontherocks.com/2019/09/cyber-war-as-an-intelligence-contest/).

comparing "digital fingerprints" of the cyber attack in question to prior attacks by known groups, and so on. However, cyber forensics remains challenging due to attackers' ability to hide their location and cover their tracks (e.g., using misdirection, providing false clues/flags), mimic the cyber aggression of other known hacking groups, and forgo custom code with known characteristics in favour of "off the shelf" tools.

Consequently, *provably* determining the identity of a hacker is difficult, which makes difficult redress via domestic justice and/or international legal mechanisms, measures for retrieving stolen data, and, in extreme cases, retaliation qua cyber self-defence. Given the possibility of a lack of retaliation by a targeted entity—and even the ***likelihood*** that a cyber exploit's expected utility will outweigh expected disutility from retaliation—the cyber domain's attribution problem brings with it an inbuilt incentive for offence (see Kyu-dok **Hong** and Seong-jong **Song** in this volume on how alliance-based retaliation could differ).

The attribution problem is implicated in the final factor supporting the superiority of cyber offence over defence: weak cyber deterrence.[30] This is obvious on the level of deterrence by punishment, as uncertainty about the identification of a cyber aggressor, as well as concern about inadvertent escalation (especially problematic for highly networked societies that would suffer from the "glass house" effect), lead the attacked target to delay, mitigate or even forgo a retaliatory response.[31] This encourages offensive cyber operations, as the attacker can frequently assume that a target's retaliation threats lack credibility.

[30] Martin Libicki (2009). *Cyberdeterrence and Cyberwar*. (Santa Monica, CA: Rand Corporation). (https://www.rand.org/content/dam/rand/pubs/monographs/2009/RAND_MG877.pdf); Timothy McKenzie (2017). "Is Cyber Deterrence Possible?" Air Force Research Institute Papers (January). (https://media.defense.gov/2017/Nov/20/2001846608/-1/-1/0/CPP_0004_MCKENZIE_CYBER_DETERRENCE.PDF). The 2020 *Foreign Affairs* article by US NSA/Cyber Command head Paul Nakasone addresses cyberspace competition without even using the word deterrence. He focuses on making cyberattacks harder and more expensive, and limiting success (more like counterespionage). See: Paul Nakasone and Michael Sulmeyer (2020). "How to Compete in Cyberspace: Cyber Command's New Approach." *Foreign Affairs* (August).

[31] Sandeep Baliga et al. (2019). "Deterrence with Imperfect Attribution." MIT Economics Department Working Paper (February 12). (http://home.uchicago.edu/~bdm/PDF/deterrence.pdf); Sandeep Baliga et al. (2019). "The Case for a Cyber Deterrence Plan that Works." *The National Interest* (March 5). (https://nationalinterest.org/feature/case-cyber-deterrence-plan-works-46207).

Deterrence by other means—denial, entanglement, and norms—also faces difficulties in dissuading attack in the cyber domain.[32] Deterrence by denial is currently a dismal failure, as evidenced by the large number of successful cyber attacks.[33] As discussed above, this is a function of the technological and cost advantages held by cyber aggressors, as well as the massive, near ubiquitous cyber vulnerability of contemporary society. Cyber attackers are insufficiently dissuaded by potential attack failure because success rates appear high. That is, deterrence by denial is failing because the positive expected utility of cyber aggression justifies it. Non-classical concepts of deterrence—entanglement and norms—are weak also. Deterrence by entanglement—which is derived from interdependence—is at best unproven in the cyber realm, and in fact it is hard to see what kinds of systemic or bilateral self-harm (the mechanism behind deterrence by entanglement) would operate to dissuade malicious cyber actors. Norm-based deterrence could work, at least in terms of creating a taboo for major cyber attacks, but precisely such international norms are lacking, and generating them is difficult, partially because it implicates putting know-how on the table, a formidable obstacle in the cyber context.

In many respects, deterrence—at least as used for military doctrine—is the wrong theoretical concept to apply to the cyber domain in order to imagine how to attenuate cyber's offensive dominance. More than an open confrontation of opposed forces, cyberspace is an intelligence challenge, a recurring competition of action and counter-action.[34] Understanding cyber in this way does not view eradication of offensive cyber operations as the only acceptable outcome, but rather accepts a certain level of ongoing competition, just as intelligence services compete with each other clandestinely. One key task therefore is for major cyber domain actors to work out—likely informally—agreed competition and manage it in order to prevent crisis instability derived from escalatory responses to offensive cyber operations.

[32] Joseph Nye (2016). "Deterrence and Dissuasion in Cyberspace." *International Security* 41(3): 44–71.

[33] Cyberspace Solarium Commission of the United State of America (2020). *Report of the Cyberspace Solarium Commission*.

[34] Rovner (2019).

4 Chainganging and Cyber Conflict Risk in East Asia

Turning to how this could play out in East Asia, preventing crisis instability and escalation in the cyber domain could be difficult because of the nature of the regional international system: offensive dominance in a multipolar environment is ripe for conflict. A classic statement of this observation is Christensen and Snyder (1990).[35] This section attempts to apply their theory of "chainganging" (over-balancing) and "buckpassing" (under-balancing) in multipolar systems to the situation of the cyber offence-defence balance in East Asia.

Christensen and Snyder (1990) show that Waltz's (1979) neorealist theory can inform us about the *aggregate* state of a multipolar international system, but cannot make determinate predictions about *individual* state policies regarding the choice to chaingang or buckpass, in the absence of hedging possibilities.[36] It is these state decisions, however, that interest policy-oriented specialists. This is relevant for a multipolar East Asia because the US, which is in a rivalry with China, operates a loose network of bilateral ("hub-and-spoke") alliances with Japan, South Korea, and the Philippines. These alliance partners will likely face difficult security policy decisions, including in the cyber domain, as it becomes harder to hedge between Washington and Beijing.[37]

Christensen's and Snyder's point is that in a multipolar order one must include perception of offence-defence balance as a variable determining whether alliance partners will chaingang or buckpass when faced with conflict stemming from an inefficient, unstable balance-of-power. Offence-defence balance perceived as overweighted to offence should produce chainganging, that is, over-balancing in which an alliance partner provides near-unconditional support for and participation in another alliance partner's conflict.[38] Offence-defence balance perceived as overweighted to defence should lead to buckpassing, that is, under-balancing in which "balancing alignments fail to form in a timely fashion because some states

[35] Christensen and Snyder (1990).

[36] Waltz (1979).

[37] Mason Richey (2019). "US-led Alliances and Contemporary International Security Disorder: Comparative Responses of the Transatlantic and Asia-Pacific Alliance Systems." *Journal of Asian Security and International Affairs* 6(3): 275–298.

[38] Robert Jervis (1978). "Cooperation Under the Security Dilemma." *World Politics* 30 (January 1978): 167–214; Christensen and Snyder (1990).

try to free ride on other states' balancing efforts."[39] Chainganging occurs when security-interdependent allies believe they are obligated (for their security interests) to support another ally's conflict (even a reckless one) because staying uninvolved risks the security of the conflict initiator such that the other alliance members (and the alliance per se) face significant security loss. In an international system displaying rising conflict risk, buckpassing happens when states fear the costs of balancing and/or predict relative power gains by free-riding on the balancing of other states (especially alliance partners). When hedging between the two options is impossible, the dilemma of choosing to chaingang or buckpass is resolved by the perceived weight of the offence-defence balance.[40] Offensive advantage incentivizes chainganging because superior offence translates into states' rapid vulnerability under attack; defensive advantage incentivizes buckpassing because even states battling alone can stalemate an aggressor.[41]

As described in Sect. 3, the cyber offence-defence balance in East Asia favours offence. Thus one should predict that the members of the hub-and-spoke alliance are likely to chaingang as conflictual situations emerge in the cyber domain. The US-China dyad is particularly worrisome, as both rivalrous states have large, well-trained cyber corps with access to many offensive cyber capabilities.

Beyond the US National Security Agency (NSA)—the largest US intelligence organization, employing more than 10,000 cyber agents and responsible for cyber-kinetic attacks such as Stuxnet—both the Central Intelligence Agency (CIA) and the US military have highly capable divisions with thousands of cyber warriors specializing in offensive operations. In the case of the US military, it has a full and independent unified combatant Cyber Command led by a four-star general. Each of the four military branches—Army, Navy, Air Force, and Marines—has service components tailored for specific operations. As for China, accurate estimates of its cyber forces are difficult, but at least 50,000 is a reasonable figure, including PLA Unit 61398 (designated as Advanced Persistent Threat [APT] 1) allegedly responsible for hundreds of defence/security cyber espionage exploits against governments (the US, South Korea,

[39] Ibid.

[40] Jervis (1978).

[41] Mason Richey (2020). "Alliance Politics and Policy in the Multipolar Indo-Asia-Pacific: The Hub-and-Spokes System Between Buckpassing and Chainganging." *The International Spectator* 55(1): 1–17.

India, Taiwan, Vietnam, inter alia), international organizations, and defence/security contractors, technology companies, and the health sector.[42]

Beijing's and Washington's cyber forces are complemented by their respective cyber postures. The US Defense Department's (DOD) *2018 Cyber Strategy and Cyber Posture Review* lists "defend[ing] forward," a euphemism for "offensive cyber operations," as one of the main employments of US cyber capabilities.[43] The strategy for accomplishing this includes "persistent engagement" with—including "preempti[on]" of—perceived cyber threats. Given the classified nature of US cyber capabilities and operational planning, it is impossible to know exactly what this entails, but a strategy intending to interdict and impose costs on adversaries for cyber attacks *before* they reach the US likely means "positioning" US cyber operations on non-US networks.[44] As Robert Chesney has written, such "positioning" could include "intrusions into the systems of potential adversaries in order to secure access ... that can be exploited for disruptive or destructive effect if and when the need later arises."[45] Depending on the context, such action may or may not violate international law, but it is undoubtedly the outgrowth of an aggressive cyber posture. Indeed, whereas the 2015 *US DOD Cyber Strategy* included risk mitigation and escalation control as important objectives, the 2018 version omits these words, instead focusing on "preemption" and "winning" in the cyber domain, including both below the threshold of armed conflict and outside military networks (i.e., inter alia, .gov and .com, rather than just .mil

[42] Mara Hvistendahl (2010). "China's Hacker Army." *Foreign Policy* (March 3). (https://foreignpolicy.com/2010/03/03/chinas-hacker-army/); David Sanger et al. (2013). "Chinese Army Unit Is Seen as Tied to Hacking Against U.S." *New York Times* (February 19).

[43] United States Department of Defense (DOD) (2018). 2018 Cyber Strategy and Cyber Posture Review; Ellen Nakashima (2018). "White House Authorizes 'Offensive Cyber Operations' to Deter Foreign Adversaries." *Washington Post* (September 20); Jeff Kosseff (2019). "The Contours of 'Defend Forward' Under International Law." 2019 11th International Conference on Cyber Conflict. (June). (https://ccdcoe.org/uploads/2019/06/Art_17_The-Contours-of-Defend-Forward.pdf).

[44] Kosseff (2019).

[45] Robert Chesney (2018). "The 2018 DOD Cyber Strategy: Understanding 'Defense Forward' in Light of the NDAA and PPD-20 Changes." *Lawfare* (September 25). (https://www.lawfareblog.com/2018-dod-cyber-strategy-understanding-defense-forward-light-ndaa-and-ppd-20-changes).

networks).[46] The US has shown political will to execute offensive cyber operations consistent with its capabilities and posture, having already conducted major cyber actions against China, Russia, North Korea, and Iran.[47]

As for Beijing, in some ways its cyber strategy and posture mirror those of Washington, contributing to a cybersecurity dilemma (see Cuihong **Cai** in this volume). To wit, China's 2015 military strategy white paper refers to both strategy and operational concepts under the rubric of "active defence," which aims at similar objectives as the US "defending forward."[48] Beyond that overriding similarity, Beijing's cyber strategy and posture present three additional challenges. First, its cyber strategy is congruent with—indeed, participates in—China's strategy in other security dimensions, notably grey-zone activities that approach and blur thresholds for sovereignty-violating hostile acts, but do not cross into areas demanding conventional military responses.[49] Second, and partially connected to the grey-zone issue (in terms of subterfuge and opacity), China's offensive cyber actions occur within the context of a "whole-of-nation, national-level strategy to 'fuse' the Chinese military and civilian industrial complexes."[50] This military-civilian fusion involves Chinese officials deploying cyber means as a part of "private" sector efforts "to persistently pursue illicit transfers of technology for military applications through …

[46] US DOD (2015). The DOD Cyber Strategy. (https://archive.defense.gov/home/features/2015/0415_cyber-strategy/final_2015_dod_cyber_strategy_for_web.pdf); Nina Kollars and Jacquelyn Schneider (2018). "Defending Forward: The 2018 Cyber Strategy is Here." *War on the Rocks* (September 20). (https://warontherocks.com/2018/09/defending-forward-the-2018-cyber-strategy-is-here/).

[47] CSIS (2019).

[48] State Council (People's Republic of China) (2015). China's Military Strategy. (http://english.www.gov.cn/archive/white_paper/2015/05/27/content_281475115610833.htm); Ben Buchanan and Robert Williams (2019). "A Deepening U.S.-China Cybersecurity Dilemma." *Lawfare* (October 24); Lyu Jinghua (2019). "What Are China's Cyber Capabilities and Intentions?" Carnegie Endowment for International Peace—IPI Global Observatory (April 1). (https://carnegieendowment.org/2019/04/01/what-are-china-s-cyber-capabilities-and-intentions-pub-78734). See also the conclusion of Kyu-dok **Hong** and Seong-jong **Song** in this volume.

[49] Kathleen Hicks et al. (2019). "China in the grey zone." European Center of Excellence Strategic Analysis 4/2019. (https://www.hybridcoe.fi/wp-content/uploads/2019/09/Strategic-Analysis-4_2019_rgb.pdf).

[50] Robert Work and Greg Grant (2019). "Beating the Americans at their Own Game: An Offset Strategy with Chinese Characteristics." CNAS Report (June 6). (https://www.cnas.org/publications/reports/beating-the-americans-at-their-own-game).

cyber interactions with the United States and other advanced states."[51] Third, China uses cyber operations as substitutes for other coercive means, especially in the context of limited conflicts.[52] Finally, China is known to be responsible for, or implicated in, hundreds of significant cyber attacks against the US, Japan, South Korea, Europe, Taiwan, and other polities over the last decade, although there is no evidence that it has executed a physically destructive attack (such as a Stuxnet-style cyber-kinetic attack).[53]

The United States and China are not the only states active in East Asia with significant cyber capabilities. Russia, North Korea, South Korea, and Japan all have significant numbers of cyber specialists and well-developed cyber postures. Russia's aggressive cyber capabilities and intent are well known.[54] Meanwhile North Korea's 7000 cyber warriors have two primary offensive tasks: (a) cyber theft (both of traditional funds and cryptocurrency) as means to partially fill hard currency shortfalls, and (b) cyber espionage (including for acquisition of intelligence and defence/security information at finance/banking, technology, military and international energy institutions).[55] Pyongyang has also carried out coercive cyber operations targeting infrastructure (such as the 2013 DarkSeoul attack), as well as ransomware attacks.[56] For its part, South Korea was late—2011—in classifying cyber as a conflict domain. Seoul first promulgated a standalone national cyber strategy in 2019, which will not be implemented before 2022. Generally South Korea's cyber posture is deterrence-oriented, as there is only limited evidence that South Korea has perpetrated

[51] Ibid.

[52] Fiona Cunningham (2018). *Maximizing leverage: explaining China's strategic force postures in limited wars.* PhD dissertation (Massachusetts Institute of Technology). (https://dspace.mit.edu/handle/1721.1/121602).

[53] Karsten Friis and Jens Ringsmose (2016). *Conflict in Cyberspace: Theoretical, strategic, and legal perspectives.* (Abingdon, UK: Routledge); Buchanan and Williams (2018); Jinghua (2019); CSIS (2019).

[54] CSIS (2019).

[55] Kim Chong Woo (2019). "The Evolution of North Korean Cyber Threats." Asan Forum (February 20). (http://en.asaninst.org/contents/the-evolution-of-north-korean-cyber-threats/); Ed Cesar (2021). "The Incredible Rise of North Korea's Hacking Army." *The New Yorker* (April 26). (https://www.newyorker.com/magazine/2021/04/26/the-incredible-rise-of-north-koreas-hacking-army).

[56] Ibid.; CSIS (2019); Quentin Hodgson et al. (2019). *Fighting Shadows in the Dark: Understanding and Countering Coercion in Cyberspace.* (Santa Monica, CA: RAND Corporation). (https://www.rand.org/pubs/research_reports/RR2961.html?utm_campaign=&utm_content=1568663851&utm_medium=rand_social&utm_source=twitter).

cyber hacks against other states or businesses.[57] Finally, as for Japan, its cyber capabilities (especially military) have lagged those of the US and China.[58] In terms of manpower alone (much less cyber tools), Japan is underdeveloped in the cyber realm, with less than 1500 cyber specialists in the military and other agencies tasked with protecting critical infrastructure.[59] This has become a weakness, as Beijing (as well as Pyongyang) has carried out cyber attacks against Tokyo with increasing frequency.[60] Correspondingly, Japan's cyber posture is defensive, with Tokyo arguing that acquisition of offensive cyber capabilities would violate its constitution.[61] Given these constraints, Tokyo is investing greater resources in growing the number of its cyber specialists, and has increased cyber cooperation with Washington in order to bolster cyber deterrence vicariously.[62]

Combining an alliance-heavy multipolar East Asia region with an offensively overweighted cyber domain and numerous states with highly capable cyber forces aggressively postured presents a cybersecurity situation with serious risks for conflict. With regional balance-of-power inherently unstable (due to multipolarity), potential crisis instability will be the default mode—especially for cyber—within a frictional context of US-China rivalry.[63] The logic is straightforward, as the tactical advantage of conflict initiation is inherent in the offensive dominance of the cyber domain, but carries with it the risk of strategic hazards (in terms of both cyber per se and the interconnection of cyber to other conflict domains).[64] Empirical reality supports this logic also, as China, in particular, and to a large extent also the US and Russia, relies on a warfighting doctrine of quickly attacking "systems and networks" including via cyber means. As Work and Grant put it:

[57] CSIS (2019).

[58] Franz-Stefan Gady (2017). "Japan's Defense Ministry Plans to Boost Number of Cyber Warriors." *The Diplomat* (July 17). (https://thediplomat.com/2017/07/japans-defense-ministry-plans-to-boost-number-of-cyber-warriors/).

[59] Gady (2017).

[60] Friis and Ringmose (2016); CSIS (2019); Export.gov (2019). Japan—Cyber Security. Export.gov (September 6). (https://www.export.gov/article?id=Japan-Cyber-Security).

[61] Gady (2017).

[62] Government of Japan (2018). Cybersecurity Strategy. (https://www.nisc.go.jp/eng/pdf/cs-senryaku2018-en.pdf); Export.gov 2019.

[63] David Gompert and Martin Libicki (2014). "Cyber Warfare and Sino-American Crisis Instability." *Survival* 56(4): 7–22.

[64] Eric Gartzke and Jon Lindsay (2017). "Thermonuclear Cyberwar." *Journal of Cybersecurity* 3(1): 37–48.

> Given its centrality to PLA strategic thought, system destruction warfare has become a dominant driver behind PLA force structure decisions and modernization priorities. It explains heavy Chinese investments in counter-battle network capabilities and the means to conduct "informationalized warfare"—the use of electronic warfare, cyber … to destroy the integrity of any U.S. battle network. For example, the Chinese have developed some type of electronic warfare capability to threaten every U.S. system and data link; one must presume they also have developed cyber-attack tools.[65]

Essentially, Beijing, Washington, and Moscow seek in early phases of conflict to use cyber operations to disrupt, disable, and/or destroy various strategically and tactically critical system nodes (including cyber systems) of their adversar(y)(ies).[66] Given the advantages of initiative in the cyber domain, the probability of escalation via chaingang scenarios is elevated. An accelerant for this escalation instability is the regional architecture of alliances (the US and its hub-and-spoke allies, as well as China and North Korea), which will be incentivized (by offence dominant cyberspace, and in particular "use-it-or-lose-it" fears) to uncritically support an alliance partner in case of cyber conflict (or other conflict that spills over into the cyber domain).

5 Policy Recommendations

Cybersecurity risk in East Asia ranges from routine cybercrime to cyber espionage to outright cyberwar. The unstable, multipolar, alliance-based security architecture in the region elevates the probability that small-scale cyber attacks metastasize and escalate into cyber conflict via chainganging. Consequently, the following cybersecurity policy measures should be considered. The overall goal is not to eliminate cyber incursions and attacks, which is unrealistic insofar as cyberspace activity so often resembles analogue spying in terms of ubiquity and tacit acceptance among states. Rather, bilateral, regional, and global governance for enhanced cybersecurity will acknowledge the character of the cyber domain while aiming at escalation prevention. Proposed measures for mitigating cyber escalation risk fall into three buckets of increasing ambition:

[65] Work and Grant (2019).
[66] Gartzke and Lindsay (2017).

5.1 *Bucket 1*

In bucket 1, the lowest end of the spectrum, are measures such as regular multilateral Track 1.5 meetings (aiming at building cyber norms), informal political agreements, and UN working group discussions. Many cyber-focused Track 1.5 meetings already occur, but the value-added of this specific bucket 1 proposal is that norm-building and exchange regarding cyber behaviour should treat it analogously to human-based intelligence collection. Output from such meetings could filter up into government, providing leaders with expectations, perspectives, and frameworks for evaluating the intent behind cyber behaviour of other states.

Another piece of low-hanging fruit is an increase in the type of limited, bilateral agreement reached by US president Barack Obama and Chinese president Xi Jinping in 2015. Their deal was limited to covering cyber theft of private company intellectual property but was considered successful (although the deal was occasionally violated by both sides).[67] Even if imperfect, such bilateral agreements could be useful, especially between cyber competitors seeking to reduce points of friction: for example, within Northeast Asia alone, China and Japan, North Korea and South Korea, North Korea and the US, the US and Russia.

The final element of bucket 1 is that the proposed UN Programme of Action[68] (a permanent UN forum following up the UN GGE and its OEWG)[69] adds to its agenda serious discussion of the acceptability of and limits to "active deterrence." The UN GGE and OEWG worked to outline priority cyber threats, discuss application of international law (and especially the UN Charter) to cyberspace, support cyber capacity building, and elaborate 11 cybersecurity norms and principles. However, the GGE

[67] Robert Farley (2018). "Did the Obama-Xi Cyber Agreement Work?" *The Diplomat* (August 11). (https://thediplomat.com/2018/08/did-the-obama-xi-cyber-agreement-work/).

[68] Geneva Internet Platform (2020). France and partners propose a programme of action for advancing responsible state behavior in cyberspace. (https://dig.watch/updates/france-and-partners-propose-programme-action-advancing-responsible-state-behaviour).

[69] United Nations Group of Government Experts on Developments in the Field of Information and Telecommunications in the Context of International Security (GGE), and Open Ended Working Group (OEWG). Its final output is a 2021 report on cyber norms, notably for hacking of critical infrastructure: United Nations (2021). Report of the Group of Governmental Experts on Advancing responsible State behavior in cyberspace in the context of international security (https://front.un-arm.org/wp-content/uploads/2021/06/final-report-2019-2021-gge-1-advance-copy.pdf).

and OEWG were vague about cyber defence, notably not endorsing a "cyber deterrence menu" (advocated by the US) and/or "anticipatory self-defence" in the cyber realm (advanced by France).[70] Given the ubiquity, persistence, seriousness, and escalation risk of cyber attacks, the international community should consider codifying—perhaps only informally—a number of proactive ("anticipatory" or "active") cyber defence measures that enhance deterrence by going beyond passive defence (basic security controls, firewalls and antivirus, patches, systems scanning and monitoring). On the conservative end of the spectrum, these enhanced deterrence and defence tactics include tarpits, sandboxes, and honeypots; hunting and beaconing; and denial and deception.[71] The more aggressive end of the spectrum includes botnet takedowns, rescue/recovery missions (for stolen information), forward defence "prepositioning" of script/access/backdoors on foreign systems for pre-emptive deterrence by denial ("hunting forward"), and "white hat" ransomware.[72] Especially the latter set of measures would require tight intergovernmental cooperation, as they may involve unauthorized access to systems and networks (although no disruption or destruction). Such cooperation would implicate communication between and among governments (and other relevant actors) in order to preclude misperception.

5.2 *Bucket 2*

Bucket 2 policy recommendations—regional multilateral cyber threat assessment and development of a regional cyber centre, such as in the EU—require greater regional cooperation, inter-regional dialogue, and transparency by officials. As with bucket 1, bucket 2 exchanges would reduce misperceptions and negative surprises by giving leaders an overview of the state-of-the-art of offensive and defensive cyber capabilities. Regional multilateral threat assessment in East Asia would ideally include

[70] United Nations (CCDCOE). 2015 UN GGE Report: Major Players Recommending Norms of Behaviour, Highlighting Aspects of International Law. (https://ccdcoe.org/incyder-articles/2015-un-gge-report-major-players-recommending-norms-of-behaviour-highlighting-aspects-of-international-law/); Geneva Internet Platform (2019). UN GGE and OEWG. (https://dig.watch/processes/un-gge).

[71] Center for Cyber and Homeland Security (2016). *Into the Gray Zone: The Private Sector and Active Defense Against Cyber Threats.* GWU CCHS Project Report (October). (https://cynergia.mx/wp-content/uploads/2016/12/CCHS-ActiveDefenseReportFINAL.pdf).

[72] Ibid.

both threat information exchange on an official regional level (supported by formal working groups), and sharing of updates on cyber posture and doctrine, as well as technical specifications about cyber tools and capabilities. Although an institution as developed and trust-based as NATO's CCDCOE (Cooperative Cyber Defence Centre) is unlikely in East Asia in the near future, a light version of the EU's Agency for Network and Information Security (ENISA) could be modelled on existing East Asian multilateral organizations, such as the Trilateral Cooperation Secretariat. An East Asian version of ENISA could disseminate reports on cyber best-practices, be a conduit for discussion of national Cybersecurity Strategies, promote cooperation during cyber emergencies, and, eventually, organize cyber crisis response exercises with regional states. A functioning East Asian version of ENISA could also interact on an inter-regional basis with ENISA, ASEAN's Cyber Capability Desk, and other regional cybersecurity institutions.

5.3 Bucket 3

Finally, bucket 3 includes robust international institutions and binding international conventions. For example, one could envision an International Cyber Court and a Convention Prohibiting (or Limiting/Restricting) Autonomous Cyber Weapons. A Cyber Court need not be as strong as the International Court of Justice, or even the International Tribunal on the Law of the Sea, but it should at minimum be independent enough that arbitral rulings on attack attribution would enjoy widespread credibility. As for possible conventions prohibiting/limiting/restricting cyber capabilities, such as autonomous weapons and code, models exist—the Ottawa Treaty on landmines, the Chemical Weapons Convention, a range of nuclear weapons–related treaties—although more widespread international cooperation at lower levels is necessary to build multilateral trust for such strong international law on cyber capabilities.

Such a regime is a long way out. Thus, a major consideration is identifying mechanisms by which best-practices, impetus, and political will for such international responses to cybersecurity dilemmas can be found. South Korea is a natural option, as it is a middle power strongly supportive of international institutions and a state with significant interest in better international cybersecurity governance. Seoul is also well connected both to important major powers (the US and China) and to other middle powers in East Asia and beyond. In this regard, the EU—with whom Seoul already holds a regular high-level Cyber Dialogue—is an especially important potential partner for South Korea in developing and propagating approaches to the challenges of cyber conflict-prevention and cybersecurity in general.

To begin with, Europe has a history of successful conception and implementation of regional confidence-building measures in the security arena, a wellspring of experience that could be applied, with Seoul as potential conduit, to East Asia with respect to cyber threat reduction and conflict-prevention. Track 1.5 dialogue between South Korea and the EU has already broached ideas that could benefit from further Seoul-Brussels road-testing before introduction into region-wide discussion. Examples include the introduction of a system of cyber "red" phones for high-level cybersecurity officials to communicate to prevent crisis escalation, and the construction of a share-able "tool-box" of best-practices along the range of cybersecurity threats.[73] For example, Brussels could share with Seoul (and beyond) the legal, technical, and strategic logic of its proposal of a regulation on Digital Operational Resilience, which aims to promote risk assessment, mitigation, system redundancy and so on for financial services firms, under the assumption that successful cyber attacks on their systems are inevitable such that resilience is a necessary objective as much as cyber defence.[74]

Beyond the aforementioned, the alignment of values and interests between South Korea and the EU opens up possibilities for norm-building in the cyber domain. One area ripe for this endeavour is the establishment of norms of cooperation between the public and private (e.g., technology industry) spheres with regard to important components of digital networks (e.g., 5G/6G gear). Public-private cooperation on norms for hardware standards—and incentives to follow them—could theoretically limit network cyber vulnerability. Seoul and Brussels should also focus on cybersecurity norm-building within the optic of cybersecurity risk as akin to intelligence (as mentioned above). Just as espionage is not generally against international law, many cybersecurity intrusions may simply be outside the realm of international legal regulation (indeed, many

[73] EU Cyber Direct (2020). EU-ROK Cyber Consultations: Resilience and Trust in Cyberspace (December). (https://www.gmfus.org/publications/eu-rok-consultations-resilience-and-trust-cyberspace).

[74] European Commission (2020). Proposal for a Regulation of the European Parliament and of the Council on Digital Operational Resilience for the Financial Sector. (https://eur-lex.europa.eu/legal-content/EN/TXT/?uri=CELEX%3A52020PC0595).

cybersecurity violations are espionage). For all that, even outside the scope of international law, espionage and intelligence collection more generally are not the Wild West: there are recognized norms (on targets, tradecraft, etc.). When violated, there are available channels of punishment (diplomatic and otherwise, including sanctions). This type of modus vivendi is imperfect, but emerged because states understood that there is no workable better alternative. Pending better international legal cooperation, something similar for cybersecurity should be welcomed, even if the unclear line between cyber *attack* and cyber *espionage* is a challenge. South Korea and the EU could be vanguard co-operators in this regard.

South Korea should prioritize scaling-up cyber cooperation with the EU to greater effect both intra- and inter-regionally, as Seoul and Brussels could potentially serve as a core for a "minilateral" caucus (including other East Asian states) providing momentum for major cyber players to increase investment in improved international cybersecurity governance. International law is currently too inchoate to solve many international cybersecurity vulnerabilities, so until that regulatory environment matures, norm-construction will (a) have to suffice and (b) require cooperative middle power actors as proponents. Why not Seoul and Brussels?

References

Sandeep Baliga et al (2019). "Deterrence with Imperfect Attribution." MIT Economics Department Working Paper (February 12). http://home.uchicago.edu/~bdm/PDF/deterrence.pdf.

Sandeep Baliga et al (2019). "The Case for a Cyber Deterrence Plan that Works." *The National Interest* (March 5).

Christopher Bartos (2016). "Cyber Weapons Are Not Created Equal." *Proceedings of the US Naval Institute* (Vol. 142, June). https://www.usni.org/magazines/proceedings/2016/june/cyber-weapons-are-not-created-equal.

Ben Buchanan and Robert Williams (2019). "A Deepening U.S.-China Cybersecurity Dilemma." *Lawfare* (October 24).

Tobias Burgers and David Farber (2021). "China's Dangerous Step Toward Cyber Conflict." *The Diplomat* (March 12). https://thediplomat.com/2021/03/chinas-dangerous-step-toward-cyber-war/.

Center for Cyber and Homeland Security (2016). *Into the Gray Zone: The Private Sector and Active Defense Against Cyber Threats.* GWU CCHS Project Report (October). https://cynergia.mx/wp-content/uploads/2016/12/CCHS-ActiveDefenseReportFINAL.pdf

Ed Cesar (2021). "The Incredible Rise of North Korea's Hacking Army." *The New Yorker* (April 26). https://www.newyorker.com/magazine/2021/04/26/the-incredible-rise-of-north-koreas-hacking-army.

Robert Chesney (2018). "The 2018 DOD Cyber Strategy: Understanding 'Defense Forward' in Light of the NDAA and PPD-20 Changes." *Lawfare* (September 25). https://www.lawfareblog.com/2018-dod-cyber-strategy-understanding-defense-forward-light-ndaa-and-ppd-20-changes.

Thomas Christensen and Jack Snyder (1990). "Chained Gangs and Passed Bucks: Predicting Alliance Patterns in Multipolarity." *International Organization* 44(2): 137–168.

Gary Corn (2019). "National Cyber Security: Navigating Gray-Zone Challenges in and through Cyberspace." In *Complex Battlespaces: The Law of Armed Conflict and the Dynamics of Modern Warfare* (eds. Winston Williams and Christopher Ford). (Oxford: Oxford University Press).

CSIS (2019). Significant Cyber Incidents. CSIS Report (September 4). https://www.csis.org/programs/technology-policy-program/significant-cyber-incidents.

CSIS (2021). Significant Cyber Incidents. CSIS Report (April 26). https://csis-website-prod.s3.amazonaws.com/s3fs-public/210326_Significant_Cyber_Events.pdf?ZKJIdGVXdQd2vXW.gFEcFQs2Ay7cDiqt.

Fiona Cunningham (2018). *Maximizing leverage: explaining China's strategic force postures in limited wars.* PhD dissertation (Massachusetts Institute of Technology). https://dspace.mit.edu/handle/1721.1/121602.

Cyberspace Solarium Commission of the United State of America (2020). *Report of the Cyberspace Solarium Commission.*

John Davis II et al (2017). *Stateless Attribution: Toward International Accountability in Cyberspace.* (Santa Monica, CA: RAND Corporation). https://www.rand.org/pubs/research_reports/RR2081.html.

Ken Dilanian and Kelly O'Donnell (2021). "Russian criminal group suspected in Colonial pipeline ransomware attack." *NBC News* (May 10). https://www.nbcnews.com/politics/national-security/russian-criminal-group-may-be-responsible-colonial-pipeline-ransomware-attack-n1266793.

Hannes Ebert and Laura Groenendaal (2020). "Cyber Resilience and Diplomacy in the Republic of Korea: Prospects for EU Cooperation." EU Cyberdirect Digital Dialogue. https://eucyberdirect.eu/wp-content/uploads/2020/08/digital-dialogue-rok.pdf.

Mark Emem (2019). "North Korea's $700 million Crypto Theft Chest Could Fund Nukes: Report." *CNN* (April 15). https://www.ccn.com/north-koreas-700-million-crypto-theft-chest-could-fund-nukes-report/.

Sandra Erwin (2018). "Defense inspector general finds key Air Force space programs vulnerable to cyber attacks, sabotage." *Space News* (August 16). https://spacenews.com/defense-inspector-general-finds-key-air-force-space-programs-vulnerable-to-cyber-attacks-sabotage/.

EU Cyber Direct (2020). EU-ROK Cyber Consultations: Resilience and Trust in Cyberspace (December). https://www.gmfus.org/publications/eu-rok-consultations-resilience-and-trust-cyberspace.

European Commission (2020). Proposal for a Regulation of the European Parliament and of the Council on Digital Operational Resilience for the Financial Sector. https://eur-lex.europa.eu/legal-content/EN/TXT/?uri=CELEX%3A52020PC0595.

Export.gov (2019). Japan—Cyber Security. Export.gov (September 6). https://www.export.gov/article?id=Japan-Cyber-Security.

Robert Farley (2018). "Did the Obama-Xi Cyber Agreement Work?" *The Diplomat* (August 11). https://thediplomat.com/2018/08/did-the-obama-xi-cyber-agreement-work/.

Federation of American Scientists. NSPMs—Donald J. Trump Administration. https://fas.org/irp/offdocs/nspm/index.html.

Karsten Friis and Jens Ringsmose (2016). *Conflict in Cyberspace: Theoretical, strategic, and legal perspectives.* (Abingdon, UK: Routledge).

Franz-Stefan Gady (2017). "Japan's Defense Ministry Plans to Boost Number of Cyber Warriors." *The Diplomat* (July 17). https://thediplomat.com/2017/07/japans-defense-ministry-plans-to-boost-number-of-cyber-warriors/.

Eric Gartzke and Jon Lindsay (2017). "Thermonuclear Cyberwar." *Journal of Cybersecurity* 3(1): 37–48.

Geneva Internet Platform (2019). UN GGE and OEWG. https://dig.watch/processes/un-gge.

Charles Glaser and Chaim Kaufmann (1998). "What is the Offense-Defense Balance and Can We Measure It?" *International Security* 22(4): 44–82.

David Gompert and Martin Libicki (2014). "Cyber Warfare and Sino-American Crisis Instability." *Survival* 56(4): 7–22.

Government of Japan (2018). Cybersecurity Strategy. https://www.nisc.go.jp/eng/pdf/cs-senryaku2018-en.pdf; Export.gov 2019.

Sue Halpern (2019). "How Cyber Weapons Are Changing the Landscape of Modern Warfare." *The New Yorker* (July 18). https://www.newyorker.com/tech/annals-of-technology/how-cyber-weapons-are-changing-the-landscape-of-modern-warfare.

Kathleen Hicks et al (2019). "China in the grey zone." European Center of Excellence Strategic Analysis 4/2019. https://www.hybridcoe.fi/wp-content/uploads/2019/09/Strategic-Analysis-4_2019_rgb.pdf.

Quentin Hodgson et al (2019). *Fighting Shadows in the Dark: Understanding and Countering Coercion in Cyberspace.* (Santa Monica, CA: RAND Corporation). https://www.rand.org/pubs/research_reports/RR2961.html?utm_campaign=&utm_content=1568663851&utm_medium=rand_social&utm_source=twitter.

Mara Hvistendahl (2010). "China's Hacker Army." *Foreign Policy* (March 3). https://foreignpolicy.com/2010/03/03/chinas-hacker-army/.

Robert Jervis (1978). "Cooperation Under the Security Dilemma." *World Politics* 30 (January 1978): 167–214.

Lyu Jinghua (2019). "What Are China's Cyber Capabilities and Intentions?" Carnegie Endowment for International Peace—IPI Global Observatory (April 1). https://carnegieendowment.org/2019/04/01/what-are-china-s-cyber-capabilities-and-intentions-pub-78734.

Meir Kalech (2019). "Cyber-attack detection in SCADA systems using temporal pattern recognition techniques." *Computers and Security* 84 (July): 225–238.

Kim Chong Woo (2019). "The Evolution of North Korean Cyber Threats." Asan Forum (February 20). http://en.asaninst.org/contents/the-evolution-of-north-korean-cyber-threats/.

Nina Kollars and Emma Moore (2019). "Every Marine A Blue-Haired Quasi-Rifleperson?" *War on the Rocks* (August 21). https://warontherocks.com/2019/08/every-marine-a-blue-haired-quasi-rifleperson/.

Nina Kollars and Jacquelyn Schneider (2018). "Defending Forward: The 2018 Cyber Strategy is Here." *War on the Rocks* (September 20). https://warontherocks.com/2018/09/defending-forward-the-2018-cyber-strategy-is-here/.

Jeff Kosseff (2019). "The Contours of 'Defend Forward' Under International Law." 2019 11th International Conference on Cyber Conflict. (June). https://ccdcoe.org/uploads/2019/06/Art_17_The-Contours-of-Defend-Forward.pdf.

Patricia Lewis and Beyza Unal (2019). "The Destabilizing Danger of Cyberattacks on Missile Systems." Chatham House—Expert Comment (April 2). https://www.chathamhouse.org/expert/comment/destabilizing-danger-cyberattacks-missile-systems.

Martin Libicki (2009). *Cyberdeterrence and Cyberwar.* (Santa Monica, CA: Rand Corporation). https://www.rand.org/content/dam/rand/pubs/monographs/2009/RAND_MG877.pdf.

Keir Lieber (2011). "Mission Impossible: Measuring the Offense-Defense Balance with Net Military Balance." *Security Studies* 20: 451–459.

Robert Lipovsky and Anton Cherapanov (2016). "BlackEnergy troan strikes again: Attacks Ukrainian electric power industry." WeLiveSecurity (January 4). https://www.welivesecurity.com/2016/01/04/blackenergy-trojan-strikes-again-attacks-ukrainian-electric-power-industry/.

Timothy McKenzie (2017). "Is Cyber Deterrence Possible?" Air Force Research Institute Papers (January). https://media.defense.gov/2017/Nov/20/2001846608/-1/-1/0/CPP_0004_MCKENZIE_CYBER_DETERRENCE.PDF.

Hannah Murphy (2019). "Cyber security companies race to combat 'deepfake' technology." *Financial Times* (August 16). https://www.ft.com/content/63cd4010-bfce-11e9-b350-db00d509634e.

Ellen Nakashima (2018). "White House Authorizes 'Offensive Cyber Operations' to Deter Foreign Adversaries." *Washington Post* (September 20).

Paul Nakasone and Michael Sulmeyer (2020). "How to Compete in Cyberspace: Cyber Command's New Approach." *Foreign Affairs* (August).

Lily Newman (2020). "No one Knows How Deep Russia's Hacking Rampage Goes." Wired (December 14). https://www.wired.com/story/russia-solarwinds-supply-chain-hack-commerce-treasury/.

Joseph Nye (2016). "Deterrence and Dissuasion in Cyberspace." *International Security* 41(3): 44–71.

Office of the United States Director of National Intelligence (2018). A Guide to Cyber Attribution. https://www.dni.gov/files/CTIIC/documents/ODNI_A_Guide_to_Cyber_Attribution.pdf.

Ankit Panda (2018). "US 'Left of Launch' Cyber Effort Might Increase Korean Peninsula Nuclear Dangers." *The Diplomat* (October, 22). https://thediplomat.com/2018/10/us-left-of-launch-cyber-efforts-might-increase-korean-peninsula-nuclear-dangers/.

Mason Richey (2019). "US-led Alliances and Contemporary International Security Disorder: Comparative Responses of the Transatlantic and Asia-Pacific Alliance Systems." *Journal of Asian Security and International Affairs* 6(3): 275–298.

Mason Richey (2020). "Alliance Politics and Policy in the Multipolar Indo-Asia-Pacific: The Hub-and-Spokes System Between Buckpassing and Chainganging." *The International Spectator* 55(1): 1–17.

Joshua Rovner (2019). "Cyber War as an Intelligence Contest." *War on the Rocks* (September 16). https://warontherocks.com/2019/09/cyber-war-as-an-intelligence-contest/.

Lora Saalman (ed.) (2019). *The Impact of Artificial Intelligence on Strategic Stability and Nuclear Risk* (Vol. II—East Asian Perspectives). SIPRI Report (October).

David Sanger et al. (2013). "Chinese Army Unit Is Seen as Tied to Hacking Against U.S." *New York Times* (February 19).

Rebecca Slayton (2017). "What Is the Cyber Offense-Defense Balance? Conceptions, Causes, and Assessment." *International Security* 41(3): 72–109.

SonicWall (2021). Cyber Threat Report. https://www.sonicwall.com/2021-cyber-threat-report/.

State Council (People's Republic of China) (2015). China's Military Strategy. http://english.www.gov.cn/archive/white_paper/2015/05/27/content_281475115610833.htm.

Jack Stubbs et al (2019). "Inside the West's failed fight against China's 'Cloud Hopper' hackers." *Reuters* (June 26). https://www.reuters.com/investigates/special-report/china-cyber-cloudhopper/.

United Nations (CCDCOE). 2015 UN GGE Report: Major Players Recommending Norms of Behaviour, Highlighting Aspects of International Law. https://ccdcoe.org/incyder-articles/2015-un-gge-report-major-players-recommending-norms-of-behaviour-highlighting-aspects-of-international-law/.

US DOD (2015). The DOD Cyber Strategy. https://archive.defense.gov/home/features/2015/0415_cyber-strategy/final_2015_dod_cyber_strategy_for_web.pdf.

United States Department of Defense (DOD) (2018). 2018 Cyber Strategy and Cyber Posture Review.

Julia Voo et al (2020). *National Cyber Power Index*. Harvard Belfer Center https://www.belfercenter.org/sites/default/files/2020-09/NCPI_2020.pdf.

Kenneth Waltz (1979). *Theory of International Politics*. (Reading, MA: Addison-Wesley).

Robert Work and Greg Grant (2019). "Beating the Americans at their Own Game: An Offset Strategy with Chinese Characteristics." CNAS Report (June 6). https://www.cnas.org/publications/reports/beating-the-americans-at-their-own-game.

Kim Zetter (2015). *Countdown to Zero Day: Stuxnet and the Launch of the World's First Digital Weapon*. (New York: Crown Publishing).

Developing a Collective Retorsion Framework Against Malicious Cyber Operations: Opportunities and Steps for EU-South Korea Cybersecurity Cooperation

Joohui Park

1 Introduction

Despite its national and international efforts to address malicious cyber operations, South Korea lacks an effective international strategy for response.[1] South Korea's experience shows at various occasions that it has

[1] In this chapter, "cyber operation" means the employment of cyber capabilities to achieve objectives in or through cyberspace, as proposed in the Tallinn Manual 2.0 on the International Law Applicable to Cyber Operations. Michael N. Schmitt (ed.), *Tallinn Manual 2.0 on the International Law Applicable to Cyber Operations*, (Cambridge University Press, 2017), 564.

J. Park (✉)
Korea University, Seoul, South Korea

© The Author(s), under exclusive license to Springer Nature Switzerland AG 2022

G. Boulet et al. (eds.), *Cybersecurity Policy in the EU and South Korea from Consultation to Action*, New Security Challenges,
https://doi.org/10.1007/978-3-031-08384-6_5

had difficulty developing feasible response options. South Korea has attributed several cyber incidents to North Korea, including the DDoS Attack against South Korean government websites in 2009, the disruption of networks of the NH Bank in 2011, and the cyber operation against South Korea's major banks and broadcasters in 2013.[2] However, Seoul has never taken any international responses. Even though Russia's military intelligence agency, GRU, was blamed by the US and the UK in October 2020 for several malicious acts including the cyber operation targeting the 2018 Pyeongchang Olympic Games, South Korea has remained silent on the topic.[3] Presumably, one of the main reasons for this inaction derives from the attribution challenges. The calculation of political costs may be another reason behind South Korea's silence.

In this context, the restrictive measures envisioned in the Framework for a Joint EU Diplomatic Response to Malicious Cyber Activities (EU Cyber Diplomacy Toolbox) should be highlighted.[4] With a special focus on the characteristics of the EU's restrictive measures as a collective retorsion which consists of lawful but unfriendly acts,[5] this chapter develops the argument that South Korea should seek cooperation with the EU to build a framework for collectively imposing proportionate costs on malicious cyber actors, a framework for collective retorsion. Section 2 underscores South Korea's two major challenges in response to malicious cyber actors: attribution challenges and high costs of unilateral countermeasures. Section 3 examines the EU Cyber Diplomacy Toolbox's restrictive measures, especially focusing on the main conditions for imposing restrictive measures. Section 4 seeks to identify possible opportunities and steps for EU-South Korea cooperation for a collective retorsion framework, by assessing the characteristics of the Toolbox's restrictive measures as collective retorsions. Section 5 gives the concluding remarks and recalls the argument this chapter makes.

[2] For a discussion on cyber threat landscape, see Mason **Richey**'s chapter.

[3] The US Department of Justice, "Six Russian GRU Officers Charged in Connection with Worldwide Deployment of Destructive Malware and Other Disruptive Actions in Cyberspace." October 19, 2020; The UK, "UK exposes series of Russian cyber attacks against Olympic and Paralympic Games," 19 October 2020.

[4] Council of the EU, *Council Conclusions on a Framework for a Joint EU Diplomatic Response to Malicious Cyber Activities ("Cyber Diplomacy Toolbox")*, Brussels, 19 June 2017.

[5] Samuele De Tomas Colatin, "Si vis cyber pacem, para sanctiones: the EU Cyber Diplomacy Toolbox in action," accessed August 24, 2021, https://ccdcoe.org/library/publications/si-vis-cyber-pacem-para-sanctiones-the-eu-cyber-diplomacy-toolbox-in-action.

2 South Korea's Challenges in Response to Malicious Cyber Operations

2.1 *Attribution Challenges*

Like for most countries, attribution is one of the main challenges South Korea is facing in response to malicious cyber operations conducted or sponsored by other states. Attribution means assigning international legal responsibility for malicious cyber operations to a state.[6] For South Korea, the first challenge in attribution occurs in collecting electronic evidence (e-evidence).[7] The nature of cyberspace—openness, global connectivity, and anonymity—allows anyone to make use of it anonymously and transnationally. Thus, malicious actors can target any information system from anywhere whenever there is a connection to the Internet. This implies that perpetrators leave digital traces in diverse jurisdictions. As a consequence, relevant e-evidence is frequently stored outside of the investigating state or by a service provider established outside its jurisdiction. South Korea is one of the most digitally connected countries. To collect e-evidence located in foreign states, however, investigative authorities of South Korea have no choice but to rely on time-consuming and resource-intensive mutual legal assistance procedures, which are traditional and outdated ways of exchanging evidence between states.[8]

To facilitate access to e-evidence stored abroad, a framework to directly obtain data from service providers can offer more effective ways to gather digital evidence.[9] The Cybercrime Convention (Budapest Convention) allows trans-border access to data directly from the person who has lawful

[6] In addition to this legal aspect of attribution, there are also different concepts on attribution: technical attribution or political attribution. The Netherlands, *Letter of 5 July 2019 from the Minister of Foreign Affairs to the President of the House of Representatives on the international legal order in cyberspace, Appendix: International law in cyberspace*, 2019.

[7] See the chapters of Gibum **Kim** and Tatiana **Tropina**.

[8] Exercise by a state of enforcement jurisdiction in a foreign state is possible under the term of a treaty (to which both states are party) or other consent of the foreign state. Generally, mutual legal assistance is governed by multilateral or bilateral treaties on mutual legal assistance in criminal matters.

[9] See the chapters of George **Christou** and Ji Soo **Lee**, Tatiana **Tropina**, and Gibum **Kim**.

authority on the data under the person's consent.[10] Furthermore, to make it easy and quick to secure and obtain e-evidence, the European Commission proposed new rules in the form of a Regulation and a Directive on 17 April 2018.[11] South Korea has not yet become a party to the Budapest Convention. Although the South Korean government is preparing for accession, failure to join the Convention may delay the process of identifying malicious actors, thereby frustrating prompt and firm attribution.[12]

Second, revealing attribution requires another careful consideration in terms of politics. In practice, even if a targeted state succeeds in identifying malicious actors with a high degree of certainty and to attribute their acts to another state under international law, the targeted state may hesitate to publicly disclose the attribution via "naming and shaming." This is because the political costs of public attribution might be too high. South Korea has experienced this kind of challenge. On 19 October 2020, the US Department of Justice accused six Russian GRU officers of several malicious cyber operations, including the operation targeting the 2018 Pyeongchang Olympic Games.[13] On the same day, the UK government confirmed the responsibility of the GRU for this operation.[14] This reaction was in stark contrast to South Korea which chose to remain silent although the effects of the operation—Internet disruptions—had occurred in the territory of South Korea.[15] Of course,

[10] Budapest Convention, Article 32(b). On 28 May 2021, the Council of Europe approved the draft for the Second Additional Protocol to the Budapest Convention on enhanced co-operation and disclosure of electronic evidence. Thomas Wahl, CoE Committee Adopts Draft on E-Evidence Protocol, accessed October 29, 2021, https://eucrim.eu/news/coe-committee-adopts-draft-on-e-evidence-protocol.

[11] European Commission, *Proposal for a Regulation of the European Parliament and of the Council on European Production and Preservation Orders for electronic evidence in criminal matters*, 17 April 2018, COM(2018) 225 final; European Commission, *Proposal for a Directive of the European Parliament and of the Council laying down harmonised rules on the appointment of legal representatives for the purpose of gathering evidence in criminal proceedings*, 17 April 2018, COM(2018) 226 final. Similarly, the US has introduced the Clarifying Lawful Overseas Use of Data Act (CLOUD Act) to speed access to electronic evidence in March 2018.

[12] Council of Europe, "T-CY News: Budapest Convention at ISCR 2019," 22–24 May 2019, https://www.coe.int/en/web/cybercrime/-/budapest-convention-at-iscr-2019.

[13] The US Department of Justice, "Six Russian."

[14] The UK, "Olympic and Paralympic Games."

[15] So Jeong Kim and Sunha Bae, "Korean Policies of Cybersecurity and Data Resilience" in *The Korean Way with Data*, (Carnegie Endowment for International Peace, 2021), 58.

whether or not to publicly attribute a malicious cyber operation to a state is entirely the sovereign decision of states. The experience of South Korea, however, shows that it may not be feasible for an individual state to publicly name and denounce the perpetrating state of malicious cyber operations, especially when they are committed by major powers like Russia or China.[16]

2.2 High Costs of Unilateral Countermeasures by South Korea

Low-intensity cyber operations—those below the threshold of an armed attack under Article 51 of the UN Charter—have become more commonplace.[17] Most of the cyber operations which South Korea has experienced, such as the DDoS Attack against South Korean government websites in 2009 and the cyber operation targeting the Pyeongchang Olympic Games in 2018, do not reach the level of an armed attack. Against malicious cyber operations at this level, the South Korean government may consider responding with countermeasures, i.e. temporary non-performance by the injured state of its international obligations towards the responsible state. However, the costs of this unilateral measure may be too high for South Korea, for the following reasons:

First, concerning the legal costs of countermeasures, it remains unclear how international law regulates acts using cyberspace. Although the third, fourth, and sixth UN Group of Governmental Experts on Developments in the Field of Information and Telecommunications (UNGGEs) have reached a consensus that existing international law—specifically, state sovereignty and international norms and principles stemming from

[16] EU Cyber Direct, "EU-ROK Cyber Consultations Resilience and Trust in Cyberspace," December 16, 2020, 2.

[17] States can exercise the right of self-defence to deter malicious cyber operations which have amounted to the level of armed attack in Article 51 of the UN Charter. See also Schmitt (ed.), *Tallinn Manual 2.0*, Rule 71. As to low-intensity cyber operations, see Michael N. Schmitt, ""Below the Threshold" Cyber Operations: The Countermeasures Response Option and International Law," *Virginia Journal of International Law*, Vol. 54, No. 3, 2014.

sovereignty—applies to the use of ICTs by States,[18] the threshold of a violation of state sovereignty by malicious cyber operations is still unclear. Thus, taking countermeasures based on self-judgement regarding the wrongfulness of malicious cyber operations may lead to the commitment of another wrongful act by South Korea under international law.[19] This is because the wrongfulness of the injured state's breach of an international obligation towards the responsible state is precluded only if the countermeasure is taken based on the precedent internationally wrongful act of the responsible state.[20]

Second, South Korea may have to shoulder high political costs for taking countermeasures, especially when taken against great powers like Russia and China.[21] Thus, South Korea's unilateral countermeasures can provoke additional retaliatory actions from Russia or China and thereby worsen South Korea's relations with them. In terms of costs-benefits calculation, costs which South Korea should pay for its unilateral countermeasures seem to be greater than the benefits it can reap from the measures. In the same vein, it is necessary to recall that the South Korean government has not taken any response against Russia in the above-mentioned cyber operations against the Pyeongchang Olympic Games in 2018.

Continuous inaction of South Korea, however, could give its potential aggressors wrong messages regarding South Korea's cyber defence capability or its will to respond. Considering the above-mentioned challenges that South Korea encounters in response to malicious cyber operations,

[18] UNGA, *Group of Governmental Experts on Developments in the Field of Information and Telecommunications in the Context of International Security*, UN Doc. A/68/98, 24 June 2013, para. 20; UNGA, *Group of Governmental Experts on Developments in the Field of Information and Telecommunications in the Context of International Security*, UN Doc. A/70/174, 22 July 2015, para. 27; UNGA, *Group of Governmental Experts on Advancing Responsible State Behaviour in Cyberspace in the Context of International Security*, UN Doc. A/76/135, 14 July 2021, para. 71(b). See also the chapter of JoonKoo **Yoo**.

[19] Every internationally wrongful act of a State entails the international responsibility of that State. There is an internationally wrongful act of a State if conduct (i) is attributable to the State and (ii) constitutes a breach of an international obligation of that State. ILC, Draft Articles on Responsibility of States for Internationally Wrongful Acts, *Report of the International Law Commission on the work of its Fifty-third session*, UN Doc. A/56/10, (23 April–1 June and 2 July–10 August 2001), Articles 1 and 2.

[20] ILC, Draft Articles, Article 22. As to the conditions for countermeasures, see ILC, Draft Articles, Articles 49–54.

[21] EU Cyber Direct, "EU-ROK Cyber Consultations," 2.

collective responses, rather than unilateral ones, could be a more realistic action. South Korea should seek a collective framework to ensure that proportionate costs for malicious acts are imposed. In this regard, the EU Cyber Diplomacy Toolbox and its restrictive measures can offer insights for South Korea's response against malicious cyber actors, as will be argued in the following sections.

3 The EU Cyber Diplomacy Toolbox: Restrictive Measures

3.1 The EU Cyber Diplomacy Toolbox

The EU Cyber Diplomacy Toolbox was adopted by the Council of the EU in June 2017.[22] The EU Cyber Diplomacy Toolbox evolved from the 2015 Council Conclusions on Cyber Diplomacy, which suggests that a common and comprehensive EU approach for cyber diplomacy could contribute to conflict prevention, mitigation of cybersecurity threats, and greater stability in international relations.[23] The 2017 Implementing Guidelines for the EU Cyber Diplomacy Toolbox clarified the Toolbox by illustrating five types of measures: preventive measures, cooperative measures, stability measures, restrictive measures, and possible EU support to member states' lawful responses.[24] These measures aim at preventing or responding to both malicious cyber operations which do constitute internationally wrongful acts and those which do not rise to the level of internationally wrongful acts but are deemed as unfriendly acts.[25] The EU Cyber Diplomacy Toolbox sets the first EU's joint diplomatic response to protect itself from malicious cyber operations from outside the EU. Its restrictive measures in particular reflect the EU's aim for more assertive measures against malicious cyber actors.

[22] See also the chapter of Michael **Reiterer.**

[23] Council of the EU, *Council Conclusions on Cyber Diplomacy*, Brussels, 11 February 2015.

[24] Council of the EU, *Implementing guidelines for the Framework on a Joint EU Diplomatic Response to Malicious Cyber Activities*, Brussels, 9 October 2017.

[25] *Implementing guidelines*, 5.

3.2 Restrictive Measures Under the EU Cyber Diplomacy Toolbox

In July 2020, the Council of the EU imposed the first-ever restrictive measures, that is, sanctions on six individuals and three entities involved in several malicious cyber operations, including those against the Organisation for the Prohibition of Chemical Weapons (OPCW) and the Ukraine power grid and those known as WannaCry, NotPetya, and Operation Cloud Hopper.[26] The six individuals were two Chinese and four Russian nationals. The three entities were located in China, North Korea, and Russia.[27] In October 2020, two more Russian individuals and one Russian entity were added to the list for cyber operations against the German federal parliament (Deutscher Bundestag) and an attempted cyber attack against the OPCW.[28]

These restrictive measures were imposed on the basis of the 2019 Council Decision on Restrictive Measures.[29] The aim of the EU restrictive measures is to deter and respond to cyber attacks with a significant effect which constitute an external threat to the Union or its member states.[30] This Council Decision implemented the EU Cyber Diplomacy Toolbox's restrictive measures and provides that restrictive measures can only be taken against cyber attacks with a significant effect which constitute an external threat to the Union or its member states.[31] More specifically, restrictive measures may be imposed under the following conditions:

First, the Council Decision on Restrictive Measures only applies to cyber attacks, defined in the Council Decision as actions involving one or more of the following: access to information systems, information system interference, data interference, or data interception.[32] Only when such

[26] *Council Decision (CFSP) 2020/1127 of 30 July 2020 amending Decision (CFSP) 2019/797 concerning restrictive measures against cyber-attacks threatening the Union or its Member States*, OJ L 246, 30.7.2020.

[27] *Council Decision (CFSP) 2020/1127.*

[28] *Council Decision (CFSP) 2020/1537 of 22 October 2020 amending Decision (CFSP) 2019/797 concerning restrictive measures against cyber-attacks threatening the Union or its Member States*, OJ L 351I, 22.10.2020. For a discussion on the cyber threat landscape, see Mason **Richey**'s chapter.

[29] *Council Decision (CFSP) 2019/797 of 17 May 2019 concerning restrictive measures against cyber-attacks threatening the Union or its Member States*, OJ L 129I, 17.5.2019, 13–19.

[30] *Council Decision on Restrictive Measures*, preamble, para. 7.

[31] *Council Decision on Restrictive Measures*, Article 1(1).

[32] *Council Decision on Restrictive Measures*, Article 1(3).

actions are not authorized by the owners or other legitimate right holders of the system or data, or not permitted under the law of the EU or the member state concerned, those actions can be deemed as cyber attacks which are subject to restrictive measures under the Council Decision.[33]

Second, the effects of cyber attacks should amount to a significant level, which will be determined on the basis of diverse elements including the scope, scale, impact, or severity of disruption caused and the amount or nature of data stolen or the scale of data breaches. For example, as to the cyber attack targeting the information system of the German Federal Parliament in April and May 2015, the EU identified the theft of a significant amount of data (16 gigabytes of data) and interference with the email accounts of several Members of Parliament as well as of Chancellor Angela Merkel.[34] The EU deemed this cyber attack as having a significant effect.[35]

Third, cyber attacks should constitute an external threat to the EU or its member states. Cyber attacks constituting external threats are those which (1) originate, or are carried out, from outside the Union, (2) use infrastructure outside the Union, (3) are carried out by any natural or legal person, entity, or body established or operating outside the Union, or (4) are carried out with the support, at the direction or under the control of any natural or legal person, entity or body operating outside the Union.[36] For example, the EU found that the cyber attacks called Operation Cloud Hopper, WannaCry, and NotPetya originated from outside the Union.[37]

Fourth, not only completed cyber attacks but also attempted ones can be subject to the restrictive measures of the EU Cyber Diplomacy Toolbox.[38] For instance, in April 2018, four Russian military intelligence officers attempted to hack into the Wi-Fi network of the OPCW in The Hague, the Netherlands. The Netherlands Defence Intelligence and Security Services disrupted the attempt and prevented possible damage to

[33] *Council Decision on Restrictive Measures*, Article 1(3).

[34] Catherine Stupp, "Germany Seeks EU Sanctions for 2015 Cyberattack on Its Parliament," *The Wall Street Journal*, June 11, 2020; M Baumgartner, *et al*, "The Breach from the East," *Der Spiegel*, 18 March 2018.

[35] *Council Decision (CFSP) 2020/1537.*

[36] *Council Decision on Restrictive Measures*, Article 1(2).

[37] See also Mason **Richey**'s Chapter.

[38] *Council Decision on Restrictive Measures*, Articles 4 and 5.

the OPCW.[39] The EU considers that if this operation had been successful, the security of the network of the OPCW would have been compromised which in turn would have hampered its ongoing investigatory work. Thus, four Russian military officers who took part in the operation were added to the restrictive measures list in July 2020.[40] In October 2020, one more Russian GRU officer and one more Russian entity were included in the list.[41]

For cyber attacks with a significant effect that constitute an external threat to the Union or its member states, two types of restrictive measures can be taken: (i) prevention of entry into or transit through the territories of the EU member states, and (ii) asset freezing. The restrictive measures do not target the state which has conducted or instructed or directed malicious cyber operations, but the persons or entities that are either directly responsible for cyber attacks, or provide financial, technical, or material support or are otherwise involved in cyber attacks.[42] This implies that the personal scope of application of the restrictive measures is quite broad. For example, Chosun Expo, known as a North Korean government front company, was listed for facilitating and providing financial, technical, or material support for several cyber attacks, including WannaCry.[43]

Given the conditions and types of restrictive measures, they can be deemed as a collective retorsion, comprised of lawful but unfriendly acts. The next section examines characteristics of restrictive measures as retorsion, and explores implications for EU-South Korea cybersecurity cooperation.

[39] The Netherlands, Ministry of Defence, *Letter to the House of Representatives regarding disruption of a GRU cyber operation in The Hague*. October 2018.

[40] *Council Decision (CFSP) 2020/1127*.

[41] *Council Decision (CFSP) 2020/1537*.

[42] *Council Decision on Restrictive Measures*, Articles 4 and 5.

[43] *Council Decision (CFSP) 2020/1127*. According to the US Department of Justice, Chosun Expo is known as a North Korean government front company. *U.S.A. v. PARK JIN HYOK*, Criminal Complaint, United States District Court for the Central District of California, June 2018, 133.

4 EU-South Korea Cooperation for Developing a Collective Retorsion Framework

4.1 Lessons from the EU's Restrictive Measures for South Korea: Collective Retorsions

The EU's framework for restrictive measures shows that retorsions can be a more effective form of response than countermeasures. The two types of restrictive measures in the Toolbox—asset freezing and travel bans—are considered retorsions. According to the Implementing Guidelines, the measures within the Toolbox, including restrictive measures, should "respect applicable international law."[44] Furthermore, asset freezing and travel bans are taken in the sovereign territories of the EU member states, thereby taking into account applicable international law and humanitarian needs.[45] Thus, restrictive measures seem to be designed for lawful but unfriendly acts of states, which are regarded as retorsions.[46]

Even though the effectiveness of the Toolbox's restrictive measures has not been proven yet, the framework for collectively imposing retorsions on malicious cyber actors provides the following lessons for South Korea's response:

First, retorsion measures, as a means of response against malicious cyber operations, provide states with more flexibility compared to countermeasures.[47] When taking retorsion measures, states do not need to examine complex legal conditions, which is required in case of taking countermeasures.[48] The only requirement for retorsions is that they should be lawful. Thus, it is not necessary to evaluate legal attribution before taking

[44] *Implementing Guidelines*, 4.

[45] *Council Decision on Restrictive Measures*, Articles 4(3), 4(6), and 5(3).

[46] Colatin, "Toolbox in action." For the concept of retorsion, see ILC, the Draft Articles, Commentary to the Chapter II (Countermeasures), para. 3.

[47] Jeff Kosseff, "Retorsion as a Response to Ongoing Malign Cyber Operations," in 2020 12th International Conference on Cyber Conflict 20/20 Vision: The Next Decade, eds. T. Jančárková, L. Lindström and M. Signoretti, I. Tolga, G. Visky (Tallinn: CCDCOE, 2020).

[48] In order to take countermeasures, states should comply with several requirements under the law of state responsibility. Most of all, it is necessary to establish whether there is an internationally wrongful act, which requires attribution of acts to a state and a breach of an international obligation of that state. For more requirements, see ILC Draft Articles, Articles 49–54.

retorsion measures.[49] As mentioned above, attribution is one of the main challenges South Korea faces in response to malicious cyber operations. In this respect, retorsion is a more feasible form of response which South Korea needs to consider.

Second, it should be highlighted that the EU's restrictive measures are imposed collectively by all EU member states. The EU underscores that clearly signalling the possible consequences of the EU's collective retorsions has a deterrent impact on the behaviour of potential malicious actors, thereby reinforcing the security of the EU and its member states.[50] Such deterrence is effective only when retorsions are imposed collectively by as many countries as possible. In other words, blocking assets in the territory of one small country alone would have little impact on malicious actors. As malicious actors are unlikely to visit the targeted state anytime soon after performing malicious acts against the targeted state, a travel ban implemented just within one country would not deter potential malicious acts. Thus, it should be highlighted that the more states join the commitment to collective retorsions, the more effective the retorsion measures are likely to be.

Third, collective retorsions such as the EU's restrictive measures can offer states opportunities to respond to malicious cyber operations while minimizing the political costs of response. Measures taken collectively by a group of states, as compared to ones taken unilaterally by an individual state, are more likely to protect each state participating in the collective measures from exacerbating their political relations with the state which is subject to the measures.[51] In addition, the Toolbox's restrictive measures

[49] Even though it is not required to establish state attribution under international law in order to take measures of retorsion, states need to identify which actors conducted malicious cyber operations.

[50] The EU Cyber Diplomacy Toolbox, Annex, para. 4. Paul Ivan, *Responding to Cyberattacks: Prospects for the EU Cyber Diplomacy Toolbox*, European Policy Centre (March 2019), 12.

[51] Annegret Bendiek, "The European Union's Foreign Policy Toolbox in International Cyber Diplomacy," *Cyber, Intelligence, and Security* 2, No. 3 (December 2018), 66; Colatin, "Toolbox in action."; Jamie Collier, "Europe's New Sanction Regime Suggests a Growing Cyber Diplomacy Presence," Fireeye, accessed April 29, 2021, https://www.fireeye.com/blog/executive-perspective/2020/08/europe-new-sanction-regime-suggest-a-growing-cyber-diplomacy-presence.html.

target specific individuals or institutions, not states.[52] In case of taking countermeasures directly against the perpetrating state, South Korea should entirely assume the consequences of such actions on its international politics and relations. Accordingly, the cost-benefit calculation of such unilateral countermeasures makes South Korea reluctant to select this option, especially against great powers like Russia and China. In contrast, participating in collective retorsions will offer South Korea a way to take action against malicious cyber actors, minimizing the deterioration of the political relations with the state which directed or instructed, or otherwise sponsored, the malicious acts.

4.2 Opportunities for and Steps Towards EU-South Korea Cooperation for a Collective Retorsion Framework

Acknowledging the strength of the EU's restrictive measures as collective retorsions, South Korea and the EU should seek cooperation to develop a new framework for imposing retorsions collectively on malicious cyber actors.

As a first step, South Korea and the EU should recognize that their cooperation to build a collective retorsion framework is advantageous to both sides. This can be supported by the cybersecurity policies promoted by the EU and South Korea as follows:

For the EU, broadening the EU's collective retorsions to non-EU members states is in line with the EU's pursuit of optimizing the global impact of the EU's restrictive measures. Indeed, the EU encourages non-EU member states to adopt a similar regime to the EU's restrictive measures. According to Article 9 of the Council Decision on Restrictive Measures, the EU shall encourage third states to adopt restrictive measures similar to those provided for in the Council Decision to maximize the impact of the restrictive measures set forth in the Council Decision.[53]

For South Korea, its cooperation with the EU on collective retorsions would be consistent with the direction of its cybersecurity strategy. In April 2019, the presidential National Security Office issued the first-ever National Cybersecurity Strategy which suggests six strategic tasks: (1) to

[52] NATO CCDCOE, "European Union establishes a sanction regime for cyber-attacks," accessed April 28, 2021, https://ccdcoe.org/library/publications/european-union-establishes-a-sanction-regime-for-cyber-attacks.

[53] *Council Decision on Restrictive Measures*, preamble, para. 7 and Article 9.

increase the safety of national critical infrastructure, (2) to enhance cyber attack response capabilities, (3) to establish governance based on trust and cooperation, (4) to build foundations for cybersecurity industry growth, (5) to foster a cybersecurity culture, and (6) to lead international cooperation for cybersecurity.[54] In September 2019, the Strategy was followed by the National Cybersecurity Basic Plan outlining concrete implementation steps for the strategic tasks of the National Cybersecurity Strategy.

In order to enhance cyber attack response capabilities, the South Korean government plans four actions: (1) to deter cyber attacks, (2) to strengthen readiness for massive cyber attacks, (3) to seek comprehensive and active means, (4) to enhance capability for countering cybercrime.[55] As to the final strategic task, the government seeks to foster bilateral and multilateral cooperation frameworks and to ensure leadership in international cooperation. To that end, South Korea is planning to deepen cybersecurity cooperation with the UN, NATO, EU, and OSCE.[56] In sum, EU-South Korea cooperation to enhance cyber attack response dovetails with South Korea's strategic tasks for cybersecurity.

As a second step, the EU and South Korea should discuss specific ways to develop a collective retorsion framework through regular dialogues. South Korea and the EU have held regular cyber dialogues since 2014. In the 6th Cyber Dialogue between South Korea and the EU in 2020, the two sides exchanged on diplomatic responses to malicious cyber operations and on possible avenues for cooperation in this area.[57] Furthermore, Track 1.5 meetings could give rise to practical and productive discussions among academia, governmental agencies, and businesses. For instance, in the Track 1.5 consultation held in October 2020, the need for enhanced cooperation between the EU and South Korea was underscored.[58]

The third step would be for the EU and South Korea to use and develop their strategic partnership. The EU-ROK strategic partnership was established based on the Framework Agreement between South Korea and the EU and its member states. Cooperation in ICT-related security aspects

[54] The ROK, *National Cybersecurity Strategy*, April 2019. See also the chapter of Kyu-dok **Hong** and Seong-jong **Song**.

[55] The ROK, *National Cybersecurity Basic Plan*, September 2019.

[56] The ROK, *National Cybersecurity Basic Plan*, 28–29.

[57] EEAS, "6th annual European Union – Republic of Korea Cyber Dialogue," 24 November2020,https://eeas.europa.eu/headquarters/headquarters-homepage/89241/6th-annual-european-union-%E2%80%93-republic-korea-cyber-dialogue_en.

[58] EU Cyber Direct, "EU-ROK Cyber Consultations."

including misuse of ICTs is covered by the EU-Korea strategic partnership.[59] Based on this, the strategic partners should coordinate positions on solving common cyber threats by setting up a collective retorsion framework.[60] In 2020, the EU-Korea strategic partnership celebrated its tenth anniversary.[61] This is the moment to bring the EU-South Korea cybersecurity cooperation to a higher level.

5 Conclusion

The peaceful use of cyberspace can be assured through a framework for ensuring that proportionate costs are paid for malicious acts in cyberspace. One good example is the EU's Cyber Diplomacy Toolbox and its restrictive measures. The EU's restrictive measures can be characterized as a collective retorsion regime under international law. In light of the main challenges which South Korea faces in response to malicious cyber operations, collective retorsions rather than unilateral countermeasures could be a more realistic and effective solution for South Korea's cybersecurity. Accordingly, South Korea should consider developing special cooperation with the EU for joint responses against malicious cyber operations which would be a win-win situation for both the EU and South Korea, given increased effectiveness and deterrence of these collective measures.

References

Baumgartner, M, et al, "The Breach from the East," *Der Spiegel*, 18 March 2018.

Bendiek, Annegret, "The European Union's Foreign Policy Toolbox in International Cyber Diplomacy," Cyber, Intelligence, and Security 2, No. 3 (December 2018).

Colatin, Samuele De Tomas, "Si vis cyber pacem, para sanctiones: the EU Cyber Diplomacy Toolbox in action," accessed August 24, 2021, https://ccdcoe.

[59] Framework Agreement, Article 15.

[60] Michael Reiterer, "The 10th anniversary of the EU-Korea Strategic partnership," *Journal of European Union Studies* 56, (October 2020): 3–20. https://brussels-school.be/sites/default/files/SpecialContribution-MichaelReiterer%28EU%EC%97%B0%EA%B5%AC%2057%ED%98%B8%29.pdf

[61] Delegation of the European Union to the Republic of Korea, EU-Republic of Korea Strategic Partnership, 30 June 2020. https://eeas.europa.eu/delegations/south-korea/81748/eu-republic-korea-strategic-partnership_en.

org/library/publications/si-vis-cyber-pacem-para-sanctiones-the-eu-cyber-diplomacy-toolbox-in-action.

Collier, Jamie, "Europe's New Sanction Regime Suggests a Growing CyberDiplomacy Presence," accessed August 24, 2021, https://www.fireeye.com/blog/executive-perspective/2020/08/europe-new-sanction-regime-suggest-a-growing-cyber-diplomacy-presence.html.

Council of the EU, *Council Conclusions on a Framework for a Joint EU Diplomatic Response to Malicious Cyber Activities ("Cyber Diplomacy Toolbox")*, Brussels, 7 June 2017a.

Council of the EU, *Council Conclusions on Cyber Diplomacy*, Brussels, 11 February 2015.

Council of Europe, "T-CY News: Budapest Convention at ISCR 2019," 22–24 May 2019, https://www.coe.int/en/web/cybercrime/-/budapest-convention-at-iscr-2019.

Council of the EU, *Implementing guidelines for the Framework on a Joint EU Diplomatic Response to Malicious Cyber Activities*, Brussels, 9 October 2017b.

Council Decision (CFSP) 2019/797 of 17 May 2019 concerning restrictive measures against cyber-attacks threatening the Union or its Member States, OJ L 129I, 17.5.2019.

Council Decision (CFSP) 2020/1127 of 30 July 2020 amending Decision (CFSP) 2019/797 concerning restrictive measures against cyber-attacks threatening the Union or its Member States, OJ L 246, 30.7.2020a.

Council Decision (CFSP) 2020/1537 of 22 October 2020 amending Decision (CFSP) 2019/797 concerning restrictive measures against cyber-attacks threatening the Union or its Member States, OJ L 351I, 22.10.2020b.

EEAS, "6th annual European Union – Republic of Korea Cyber Dialogue," 24 November 2020, https://eeas.europa.eu/headquarters/headquarters-homepage/89241/6th-annual-european-union-%E2%80%93-republic-korea-cyber-dialogue_en.

EU Cyber Direct, "EU-ROK Cyber Consultations Resilience and Trust in Cyberspace," December 16 2020.

European Commission, *Proposal for a Regulation of the European Parliament and of the Council on European Production and Preservation Orders for electronic evidence in criminal matters*, 17 April 2018, COM(2018) 225 final.

European Commission, *Proposal for a Directive of the European Parliament and of the Council laying down harmonised rules on the appointment of legal representatives for the purpose of gathering evidence in criminal proceedings*, 17 April 2018, COM(2018) 226 final.

ILC, Draft Articles on Responsibility of States for Internationally Wrongful Acts, *Report of the International Law Commission on the work of its Fifty-third session*, UN Doc. A/56/10, (23 April–1 June and 2 July–10 August 2001).

Ivan, Paul, *Responding to Cyberattacks: Prospects for the EU Cyber Diplomacy Toolbox*, European Policy Centre (March 2019).

Kim, So Jeong and Bae, Sunha, "Korean Policies of Cybersecurity and Data Resilience" in *The Korean Way with Data*, (Carnegie Endowment for International Peace, 2021).

Kosseff, Jeff "Retorsion as a Response to Ongoing Malign Cyber Operations," in 2020 12th International Conference on Cyber Conflict 20/20 Vision: The Next Decade, eds. T. Jančárková, L. Lindström and M. Signoretti, I. Tolga, G. Visky (Tallinn: CCDCOE, 2020).

NATO CCDCOE, "European Union establishes a sanction regime for cyber-attacks," accessed April 28, 2021, https://ccdcoe.org/library/publications/european-union-establishes-a-sanction-regime-for-cyber-attacks.

Reiterer, Michael, "The 10th anniversary of the EU-Korea Strategic partnership," *Journal of European Union Studies* 56, (October 2020).

Schmitt, Michael N. (ed.), *Tallinn Manual 2.0 on the International Law Applicable to Cyber Operations*, (Cambridge University Press, 2017).

Schmitt, Michael N, ""Below the Threshold" Cyber Operations: The Countermeasures Response Option and International Law," *Virginia Journal of International Law*, Vol. 54, No. 3, 2014.

Stupp, Catherine, "Germany Seeks EU Sanctions for 2015 Cyberattack on Its Parliament," *The Wall Street Journal*, June 11, 2020.

The Netherlands, Ministry of Defence, *Letter to the House of Representatives regarding disruption of a GRU cyber operation in The Hague*. October 2018.

The Netherlands, *Letter of 5 July 2019 from the Minister of Foreign Affairs to the President of the House of Representatives on the international legal order in cyber-space, Appendix: International law in cyberspace*, 2019.

The ROK, *National Cybersecurity Strategy*, April 2019a.

The ROK, *National Cybersecurity Basic Plan*, September, 2019b.

The UK, "UK exposes series of Russian cyber attacks against Olympic and Paralympic Games," 19 October 2020.

The US Department of Justice, "Six Russian GRU Officers Charged in Connection with Worldwide Deployment of Destructive Malware and Other Disruptive Actions in Cyberspace," October 19, 2020.

UNGA, *Group of Governmental Experts on Developments in the Field of Information and Telecommunications in the Context of International Security*, UN Doc. A/68/98, 24 June 2013.

UNGA, *Group of Governmental Experts on Developments in the Field of Information and Telecommunications in the Context of International Security*, UN Doc. A/70/174, 22 July 2015.

UNGA, *Group of Governmental Experts on Advancing Responsible State Behaviour in Cyberspace in the Context of International Security*, UN Doc. A/76/135, 14 July 2021.

Wahl, Thomas, CoE Committee Adopts Draft on E-Evidence Protocol, accessed October 29, 2021, https://eucrim.eu/news/coe-committee-adopts-draft-on-e-evidence-protocol.

PART II

Comparisons

Cyber Arms Control on the Korean Peninsula: Challenges and Opportunities with a View to North Korea's Asymmetry Strategy

Sangho Lee and Maximilian Ernst

1 Introduction

Cyber-enabled technologies permeate all areas of society and industry today. Defence technologies are no exception to this. But the scope of cybersecurity in the inter-state context goes even further, as also domains such as information warfare, state-sponsored terrorism, and criminal activities crossing state borders, including the hacking of financial institutions and cryptocurrency exchanges, are enabled by a country's cyberwarfare capabilities. Despite the proliferation of cyber-enabled technologies, and

S. Lee
Daejeon University, Daejeon, South Korea

M. Ernst (✉)
Vrije Universiteit Brussel (VUB), Brussels, Belgium
e-mail: maximilian.ernst@vub.be

© The Author(s), under exclusive license to Springer Nature Switzerland AG 2022

G. Boulet et al. (eds.), *Cybersecurity Policy in the EU and South Korea from Consultation to Action*, New Security Challenges,
https://doi.org/10.1007/978-3-031-08384-6_6

although cyber attacks happen daily, little progress in terms of international-legally binding agreements on the use of cybercrime and warfare has been made, let alone in terms of cyber arms control.

Given the ubiquitous occurrences of cyber attacks, including cybercrime and cyberterror, on the Korean Peninsula, and in light of the possibility of an escalating conflict in which cyber capabilities would most certainly be deployed to assist kinetic attacks, an agreement regulating the use of cyber-enabled technologies has considerable potential. Without an agreement on cybersecurity, an inter-Korean peace and reconciliation process will not be successful. Without a legal framework and regulations, the involved parties have no common language to address differences and disputes in the cyber realm. Even a minor dispute resulting from a cyber attack has the potential to escalate and perturbate any hopes of peace and reconciliation between the two Koreas.

This chapter provides an overview of the challenges that result from North Korea's cyber aggressions targeting South Korea to this date and outlines opportunities for an inter-Korean cyber agreement based on the concept of arms control. Section 2 will outline the current cybersecurity situation on the Korean Peninsula in the context of the protracted Korean conflict. First, a brief historic overview of North Korea's asymmetry strategy and contextualization of the role of North Korea's cyber capabilities will be provided. Second, North Korea's use of cybercrime and cyber terror against South Korea during peacetime will be illustrated, followed by an assessment of North Korea's use of cyber weapons in case of conflict on the Korean Peninsula. Section 3 then reviews a number of existing arms control agreements and assesses their applicability to the cyber realm. Section 4 concludes with a discussion of the role that cyber arms control can play to enhance stability on the Korean Peninsula.

2 The Inner-Korean Conflict and the Role of Cyber-Enabled Strategies

2.1 *Historical Context of North Korea's Asymmetry Strategy and the Role of Cyber Capabilities*

Since 1953, the Korean Peninsula has been locked in a Cold War which did not cease to persist even when the Iron Curtain fell in Europe in the aftermath of the collapse of the Soviet Union in the early 1990s. The two

Koreas have, since the armistice of July 1953, taken diverging paths and developed differently. South Korea has transformed into a vibrant democracy and one of the world's largest economies, boasting internationally competitive industries with global export markets. North Korea, on the other hand, has suffered under decades of dictatorship, a planned economy, and protracted famines in the 1990s—all factors which have profoundly inhibited economic growth and development. This resulted in two strongly differing societies and state structures which are reflected also in the military organization and ensuing doctrines. Whereas South Korea was able to invest in cutting-edge military technology, North Korea's conventional military capability has deteriorated over the past decades.[1] The dire state of affairs of North Korea's conventional forces has further been exacerbated by food shortages and limited access to fuel due to international sanctions.

With the end of the Cold War, North Korea lost in the Soviet Union its closest ally. Pyongyang saw itself confronted with an ironclad US-ROK alliance and an altogether strong US military presence in the region. Whereas the 1961 "Sino-North Korean Mutual Aid and Cooperation Friendship Treaty" provided North Korea with comparatively limited security assurance from China, the US extended conventional and nuclear deterrence to its treaty allies Japan and South Korea. This provided a formidable and stable security environment which in turn allowed economic development. Unable to compete economically and with conventional military strength, North Korea opted to pursue an asymmetry strategy in exploiting weaknesses of the US-ROK alliance and invest first and foremost in nuclear weapons, missile technologies, and offensive cyber capabilities. The latter serve as a weapon to attack essential military and civilian infrastructure as part of national defence, as means to earn money through cybercrime, and as a means for propaganda and information warfare.

Asymmetric warfare is based on a strategy that acknowledges and makes use of the adversary's capacity for error and delays.[2] Cyber capabilities are the weapons systems of choice for states with inferior conventional military power; for a discussion on the offence-defence (im)balance in

[1] Jin Moo Kim, "The Reality of North Korea's Conventional Power: Transition to Asymmetric Strategy," in *North Korea Conundrum*, ed. Un-Chul Yang (Seongnam: Sejong Institute, 2019), 55.

[2] Peter Rosen, "Competitive Strategies: Theoretical Foundations, Limits, and Extensions," in *Competitive Strategies for the 21st Century: Theory, History, and Practice*, ed. Thomas G. Mahnken (Stanford: Stanford University Press, 2012), 18-19.

cyberconflict, see also the chapter by Mason **Richey**. Today, most modern armies possess so-called networked militaries, which integrate various sensors and systems, such as radars and missile defence. Due to its limited resources, North Korea cannot invest in conventional military capabilities to compete with its adversaries. The more economical alternative for North Korea is to pursue an asymmetry strategy and to disrupt and neutralize its adversaries' advanced conventional systems.[3]

At the core of an asymmetry strategy in cyberspace is the incapacitation of the enemy's capabilities by disrupting computer systems through malware or the manipulation of enemy computer networks and related facilities and military systems.[4] But cyberspace is not only to be understood as a military domain of its own. It also permeates into the existing domains of warfare on air, land, sea, space; it covers command and control, situational awareness, and targeting, as they are technologically advanced and based on computer systems.[5]

North Korea's asymmetry strategy is based on two pillars: first, the nuclear arsenal as a strategic deterrent that can ensure the regime's safety and thereby grants the leadership in Pyongyang legitimacy; second, cyber capabilities, which enable the country to retain the strategic initiative and to coerce its adversaries below the level of triggering nuclear war.[6] This is in line with North Korea's *Byungjin* doctrine, pronounced in 2013 by Kim Jong-Un. Thereby North Korea will simultaneously pursue economic development as well as nuclear weapons.[7] Reviews of various indicators of North Korea's economic development, including GDP growth estimates and foreign trade data, suggest that North Korea's economy continued moderate growth despite sanctions since Kim Jong-un's assumption of

[3] Jenny Jun, Scott LaFoy, and Ethan Sohn, "North Korea's Cyber Operations: Strategy and Responses" (Washington, DC: Center for Strategic & International Studies, 2015), 24.

[4] Dal Hyung Bae, "North Korea's Cyber Threat and Options for South Korea's Policy Improvement and Capability Enhancement" (Korea Research Institute of Strategy Policy, Seoul, 2011), 7-8.

[5] Marcus Willett, "Assessing Cyber Power," *Survival* 61, no. 1 (January 2, 2019) https://doi.org/10.1080/00396338.2019.1569895, 87.

[6] Maximilian Ernst and Sangho Lee, "Countering Cyber Asymmetry on the Korean Peninsula: South Korea's Defense Against Cyber Attacks from Authoritarian States," *Journal for Intelligence, Propaganda and Security Studies* 15, no. 1 (2021): 165–79.

[7] Kim, "The Reality of North Korea's Conventional Power: Transition to Asymmetric Strategy," 63.

power,[8] until the North Korean leadership in 2020 disconnected the country from the outside world entirely to prevent covid-19 imports.[9] In 2016 and 2017, North Korea conducted numerous ballistic missile and nuclear tests and announced, after successfully testing the ICBM *Hwasong 15*, that it secured a full nuclear strike capability. This means that North Korea is plausibly able to deliver a nuclear warhead to US homeland, although some doubts remain since the re-entry vehicle has not yet been tested.[10]

The cost-benefit ratio of cyber attack capabilities outweighs that of long-range ballistic missiles and nuclear weapons. While North Korea's nuclear and long-range missile tests reportedly have cost about 900 million USD, North Korea's cyber capabilities require less investment and are even used to generate income, as the following section will demonstrate. In sum, cyberwarfare for North Korea is an effective means to reinforce its asymmetry strategy on both the military-strategic and financial-budgetary fronts.[11] Cyberwarfare offers the North Korean leadership a stealthy and destructive capability that fittingly complements its nuclear arsenal.

Cyber capabilities serve another function in North Korea's strategic calculus; North Korea frequently uses locally contained conventional provocations to create or keep tensions high. If, however, the US and South Korea decided to retaliate by conventional means, the North Korean leadership would find itself out of options to respond, short of threatening nuclear war, a far too risky response. The use of cyber attacks against South Korea gives North Korea strategic benefits as it can execute these attacks at will without using significant resources. Whereas the US-ROK alliance has to endure North Korea's inconspicuous threats and attacks, Pyongyang can establish tensions on the Peninsula similar to the Cold War, which strengthens the leadership's legitimacy and grip on power. Through a continuation of such small-scale challenges directed at

[8] Maximilian Ernst and Eliana Kim, "Economic Development Under Kim Jong-Un: The Added Value of Traffic Data and Established Indicators in the Study of North Korea's Economy," *North Korean Review* 16, no. 2 (2020): 26–48.

[9] Mitch Shin, "What Is the Truth About COVID-19 in North Korea?," January 6, 2021, https://thediplomat.com/2021/01/what-is-the-truth-about-covid-19-in-north-korea/.

[10] An overview of North Korea's ballistic missile arsenal is provided by: CSIS, "Missiles of North Korea," Missile Defense Project (Washington, DC: Center for Strategic & International Studies, June 14, 2018), https://missilethreat.csis.org/country/dprk/.

[11] "What Did N.Korea's Nuclear Program Cost?", *The Chosun Ilbo*, February 13, 2013, http://english.chosun.com/site/data/html_dir/2013/02/13/2013021301335.html.

its southern neighbour, Pyongyang can continue its political onslaught, retain the strategic initiative, and maintain the legitimacy of Kim Jong-un's rule.

2.2 *North Korea's Cybercrime and Cyberterror Against South Korea in Peacetime*

Over the past two decades, the North Korean leadership has managed to allocate resources to develop cyber capabilities despite considerable economic hardship and a rigorous UN sanctions regime. It requires resource-intensive research and development over many years to establish cyber units of the quality that North Korea has. The Pyongyang leadership decided to provide its cyber programme with budgetary backing, knowing that cyber capabilities would increase its asymmetric capacity *vis-à-vis* its adversaries. Hence, North Korea has established several dedicated cyber-warfare units, which are discussed in some detail by Kyu-dok **Hong** and Seong-jong **Song** in this volume.

North Korea's cyber attacks have become more advanced and audacious in recent years. The hacker group Lazarus extracted 81 million USD from a Bank in Bangladesh in 2016 and 10 million USD from the Chilean National Bank in 2019. Lazarus, also known as "APT (Advanced Persistent Threat)-38," was identified by the leading cybersecurity company *Fire Eye* to be sponsored by North Korea.[12] The same group is suspected of the cryptocurrency hacking that resulted in the loss of over 500 million USD, targeting mainly South Korean exchanges (*Bithumb*, *Coinis*, *Youbit*, *Yapizon*) but also one exchange from Japan (*Coincheck*) throughout 2017 and 2018.[13]

North Korea also engages in acts of cyberterror. The concept of cyberterror may vary depending on the context and the actions that are taken. Based on the FBI's definition, "cyberterrorism" consists of a

[12] Nalani Fraser et al., "APT38: Details on New North Korean Regime-Backed Threat Group" (FireEye, October 3, 2018), https://www.fireeye.com/blog/threat-research/2018/10/apt38-details-on-new-north-korean-regime-backed-threat-group.html.

[13] Daniel Palmer, "North Korean Hacking Group Lazarus Stole $571 Million in Cryptos: Report" (Coindesk, October 19, 2018), https://www.coindesk.com/north-korean-hacking-group-lazarus-stole-571-million-in-cryptos-report; Neil Mathew, "$571 Million: Notorious North Korean Hacker Group Has Stolen a Fortune in Cryptocurrency" (CCN, October 19, 2018), https://www.ccn.com/571-million-notorious-north-korean-hacker-group-has-stolen-a-fortune-in-cryptocurrency/.

"premeditated, politically motivated attack against information, computer systems, computer programs and data which results in violence against non-combatant targets by subnational groups or clandestine agents."[14] Cyberterror can be launched to propagate the attacker's political and ideological beliefs by formulating threats and creating panic in the population or government, and by disrupting public infrastructure, such as railways, electricity, hospitals, or banks.[15]

Pyongyang has a vested interest in engaging in cyberterror during peacetime. This is not just aimed at causing disruptions to the public service and financial markets, but also triggering public disillusionment with the government's inability to counter and avert such attacks. North Korea has engaged in cyberterror activities aimed at undermining the authority or credibility of government officials in South Korea. Employing disinformation, slur, and character assassination, it attempted to influence or neutralize the South Korean government. The dissemination of fake news and the manipulation of media and public opinion also fall under this category which enables the attacker to destabilize the target country. Most notably, this tactic does not require hacking South Korean government or civilian computer systems. Instead, North Korea was able to collect personal data and create profiles of about 1,650,000 South Korean citizens through utilizing CRM (customer relationship management) campaigns as well as collecting data from SNS (social networking services).[16] Personalized data on citizens allow profiling and to launch customized campaigns to influence elections and to create distrust in society.[17]

South Korea is particularly vulnerable to cyberterror for two main reasons: First, South Korea is a highly integrated digital society, frequently connected to Internet services with many devices. Second, South Korea's Internet is hyper-localized, where the majority of users predominantly use

[14] Peter W. Singer, "The Cyber Terror Bogeyman" (Brookings, November 1, 2012), https://www.brookings.edu/articles/the-cyber-terror-bogeyman/.

[15] Sangho Lee, "The Realities of Cyber War and Measures to Counter Future Cyberattacks (in Korean)," *Shidae Jungshin*, Autumn 2009.

[16] Sangho Lee et al., "A Conceptual Study on Knowledge-Based Personal Information Gathering Methods for Conducting Psychological Warfare in Cyberspace," in *Convergence and Hybrid Information Technology*, ed. Geuk Lee et al., vol. 310 (Berlin, Heidelberg: Springer Berlin Heidelberg, 2012), 287–93, https://doi.org/10.1007/978-3-642-32692-9_37.

[17] Roberto J. González, "Hacking the Citizenry?: Personality Profiling, 'Big Data' and the Election of Donald Trump," *Anthropology Today* 33, no. 3 (June 2017): 9–12, https://doi.org/10.1111/1467-8322.12348

local, that is, Korean language software and Internet services. For example, *Hangul*, a Korean word processor, has many vulnerabilities that were exploited by North Korean hackers to infiltrate networks. In 2014, North Korea was able to hack Korea Hydro Nuclear Power (KHNP), a subsidiary of the Korea Energy and Power Corporation (KEPCO), using this vector.[18]

Attribution of cyber attacks has proven easier than many believe since coding patterns allow forensic detection of the perpetrator. In the cases of cyber attacks that North Korea launched using *Hangul*, the hackers used coding patterns that were already previously utilized by their organization.[19] Contrary to the often-indicated problem of attribution in cyberspace, most major cyber attacks have been attributed.[20]

2.3 North Korea's Cyber Strategy During Open Military Conflict

Within North Korea's armed forces, cyberwar units are disproportionately strong. This is because North Korea's conventional forces overall are in bad shape and because North Korean strategists began to invest in cyberwarfare capabilities early on. To date, North Korea's cyberwarfare forces are estimated to include 6000 personnel who are divided across four organizations: the General Reconnaissance Department, the International Liaison Department, the United Front Department, and the Central Party Examination Department.[21]

North Korea's focus so far has been to identify South Korean cyber vulnerabilities and to prepare to exploit them. North Korea has executed cyber espionage operations against South Korean military and strategic assets. In 2013 North Korea succeeded in hacking crucial systems, including the Aegis Air Defence system's source codes, details of the ROK armed forces' joint tactical data links as well as technical parameters for

[18] Dylan Stent, "Can the Cyber Terrorism Tactics of North Korea Be Deterred?", *The National Interest*, September 23, 2018, https://nationalinterest.org/blog/skeptics/can-cyber-terrorism-tactics-north-korea-be-deterred-31662.

[19] Lee, "The Impact of Advanced-Technology Cyber Weapon on the Future Conflict in the Korean Peninsula and East Asia," 88.

[20] Thomas Rid and Ben Buchanan, "Attributing Cyber Attacks," *Journal of Strategic Attacks* 38 (n.d.): 1–37.

[21] Seong Kyu Ahn, "Cyber Cancer: Can We Prepare for APT (Advanced Persistent Threat)," Column (Seoul: Asan Institute, August 5, 2015).

Identification of Friendly or Foe (IFF) equipment.[22] In 2016, North Korean hackers succeeded in infiltrating South Korea's military cyber command. These incidents show North Korean cyberwarfare units' ability to penetrate ROK military systems. North Korea was likely able to collect intelligence on further vulnerabilities of South Korean networks and systems and planted APTs. [23]

North Korean cyberwarfare units can also exploit civil contractors working for the armed forces. They are employed for logistics and network maintenance. As North Korean intelligence services operate sizeable human assets in South Korea, it can infiltrate the supply chains of these civil contractors and thereby embed malware in critical systems, such as South Korea's defence network.[24] North Korea is also expected to install "logic bombs," similar to Stuxnet, in critical ROK networks, which initiate autonomously once certain conditions are met.[25] Provided that North Korean cyber forces can choose the timing of a crisis in advance, it will be well prepared to launch substantial cyber attacks against South Korea with APTs and logic bombs.[26] The most likely military target for North Korean cyberwar units is the ROK's Army Tactical Command Information System (ATCIS). ATCIS is a C4I (command, control, communications, computer, and intelligence) network designed to automatically combine data from tactical military units and share them in real-time.[27]

[22] Seong Un Yoo, "Aegis, KF-16 Joint Operations and Core Secrets Were Hacked (in Korean)," *Joongang Ilbo*, September 10, 2015, https://news.joins.com/article/18631228.

[23] Lee, "The Impact of Advanced-Technology Cyber Weapon on the Future Conflict in the Korean Peninsula and East Asia," 97.

[24] Ibid, 98.

[25] John Richardson, "Stuxnet as Cyberwarfare: Applying the Law of War to the Virtual Battlefield," *John Marshall Journal of Computer and Information Law* 29, no. 1 (2011): 1–28.

[26] Brandon Valeriano and Rayn Maness, Cyber War versus Cyber Realities: Cyber Conflict in the International System (Oxford: Oxford University Press, 2015), 25.

[27] Lee, "The Impact of Advanced-Technology Cyber Weapon on the Future Conflict in the Korean Peninsula and East Asia," 99.

3 Towards an Inter-Korean Cyber Arms Control Agreement

Agreements to improve cybersecurity are a relatively new field, with virtually no track record of successful cases in the past.[28] One possible way to address the challenges and opportunities of a cyber agreement is to approach the issue as one of arms control. Arms control has been a practice with considerable success over the past decades, and challenges posed by new technologies, such as nuclear and chemical weapons, could be addressed and resolved to a significant degree.[29] Maybaum and Tölle define a cyber weapon as "a weapon that would comprise any computer equipment or device that is designed, intended or used to have violent consequences; that is, to cause death or injury to persons or damage or destruction of objects." They further define a cyber arm as "a cyber weapon that is used in a computer network."[30]

The literature largely identifies two types of arms control, namely "prevention" and "regulation." Prevention-oriented arms control mechanisms have so far been applied to situations where "a new threat or capacity has been identified, but has not yet been realised or is judged peripheral to the actual security requirements of a state."[31] An example is the 1967 Outer Space Treaty, which prohibited the placement of weapons of mass destruction and the militarization of outer space and the moon. Regulation-oriented arms control mechanisms have "tended to be utilised when certain weapons systems have already been incorporated into state arsenals and military doctrines, but where a mutual interest exists in controlling

[28] Prominent failed cases include the 2016 US-China cyber agreement: John W. Rollins, "U.S.-China Cyber Agreement" (Federation of American Scientists, October 16, 2015), https://fas.org/sgp/crs/row/IN10376.pdf.

[29] Sean M. Lynn-Jones, "A Quiet Success for Arms Control: Preventing Incidents at Sea," *International Security* 9, no. 4 (1985): 154–84; April Carter, *Success and Failure in Arms Control Negotiations* (Oxford: Oxford University Press, 1989); Matthew Fuhrmann and Yonatan Lupu, "Do Arms Control Treaties Work? Assessing the Effectiveness of the Nuclear Nonproliferation Treaty," *International Studies Quarterly* 60, no. 3 (September 2016): 530–39, https://doi.org/10.1093/isq/sqw013.

[30] Markus Maybaum and Jens Tölle, "Arms Control in Cyberspace – Architecture for a Trust-Based Implementation Framework Based on Conventional Arms Control Methods," in *Cyber Power* (8th International Conference on Cyber Conflict, NATO, 2016), 161.

[31] Paul Meyer, "Cyber-Security through Arms Control: An Approach to International Co-Operation," *The RUSI Journal* 156, no. 2 (April 2011): 25, https://doi.org/10.1080/03071847.2011.576471.

destabilising increases or deployments."[32] Examples are the strategic nuclear arms reduction agreements between the US and the Soviet Union and the multilateral Conventional Forces in Europe (CFE) Treaty. It is also possible to combine preventive and regulatory elements into one agreement, as was the case for the 1972 Anti-Ballistic Missile Treaty.[33]

Maybaum and Tölle reviewed four successful arms control regimes of the past decades and analysed what lessons to draw from them for future cyber arms control regimes. The four treaties reviewed were the CFE, the Vienna Document (VD), the Open Skies Treaty (OST), and the Chemical Weapons Convention (CWC). For the CFE and the VD, both signed in 1990, on-site inspection was a key feature, which does not apply to cyberspace. However, new technological developments in machine learning enable the detection of cyber weapons, which could serve as an equivalent screening method.[34] The CFE and VD were very useful in identifying so-called treaty limited equipment, for example, "defining exact categories of weapons systems for which the treaty was supposed to be applicable."[35] Furthermore, invitations of other armed forces to military exercises have played an important role as part of confidence-building measures. Having observers present, either physically or virtually, during tests of cyber weapons or demonstrations of cyber defence, could be a meaningful option to build trust also in the cyber context.

The OST, signed in 1992 and ratified in 2002 between NATO and Russia, enabled observation flights over the entire territory of all parties. The flights had to be notified three days in advance, and the sensors of the aircraft had specifically defined parameters to make possible the identification of treaty limited equipment. The OST worked as a tool for other verification regimes, as it had no treaty limited equipment of its own. The territorial scope of the treaty applied to cyberspace is difficult to enforce. The problem is that software and malicious code can easily be hidden or transferred to servers outside of a country. The main lesson from the OST for cyber arms control is rather that it served as a model for a cooperative approach among experts of various countries who participated in

[32] Ibid, 25.

[33] Ibid., 25.

[34] Eddie Segal, "The Impact of AI on Cybersecurity" (IEEE Computer Society, n.d.), https://www.computer.org/publications/tech-news/trends/the-impact-of-ai-on-cybersecurity.

[35] Maybaum and Tölle, "Arms Control in Cyberspace," 162.

overflights. It instilled trust and a sense of cooperation between the treaty parties, their military personnel, and defence industry experts.

The 1993 CWC "aims to eliminate all existing chemical weapons globally by prohibiting the development, production, acquisition, stockpiling, retention, transfer, or use" of chemical weapons by state parties.[36] Under the Convention, it is clearly defined what chemical agents fall under the scope of the treaty, and each party, upon ratification, negotiates an action plan with the Organisation for the Prohibition of Chemical Weapons (OPCW) how to implement the Convention and dispose of all chemical weapons it possesses. The Convention has a global scope and only four states (Egypt, Israel, North Korea, and South Sudan) have not ratified it. Furthermore, the Convention includes inspections by the OPCW, which are triggered when a convention member accuses another member of having breached the terms of the Convention. The CWC possesses useful lessons for cyber arms control. As the chemical materials are difficult to identify, it focuses instead on the specific payload of the weapon and the impact it would have if deployed. This applies to cyber weapons, where the malicious function may be undetectable until the weapon is launched, but the impact of a cyber weapon can be described in predefined metrics, for example as a breach of the integrity of a system or limitations of its availability (Table 1).[37]

Table 1 Selected arms control agreements and their applicability to cyberspace

Historical treaty	*Type of control*	*Applicability/lesson for cyberspace*
1990 Conventional Forces Europe (CFE) Treaty	Prevention	On-site inspection, to be replaced by machine-learning-enabled screening mechanisms.
1990 Vienna Document (VD)	Regulation	Ibid.
1992 Open Skies Treaty (OST)	Regulation	Bring together cybersecurity experts from various countries, instil trust and a sense of cooperation between military personnel and experts of participating countries.
Chemical Weapons Convention (CWC)	Prevention	Define cyber weapons by their potential impact. Strong enforcement mechanism; wide membership.

[36] Ibid, 165.
[37] Ibid.

All successful arms control agreements of the past have a common property; the political will of the respective member state governments to mitigate the threat posed by certain weapons technologies to their populations. The political will can stem from the desire to build stable relations and decrease the likelihood of conflict, as seen between Russia and NATO after the Cold War, or from the general appreciation that a certain weapons system or technology is especially cruel or dangerous, which was the case for nuclear and chemical weapons.

The two Koreas need to understand that cyberwar and cyberterror have the potential to jeopardize the status quo and render any peace and reconciliation outlook impossible if there were significant damage due to cyber attacks. Since attribution is possible in most cases, each perpetrator needs to be aware that every attack will potentially lead to a diplomatic, in-kind, or even military retaliation. A cyber arms control agreement could include a code of conduct between the two Koreas that clearly defines which actions in cyberspace are permissible and which are prohibited. It would also need an enforcement mechanism, based on penalties, and an enforcing organization, such as the OPCW under the CWC. In her chapter, Cuihong **Cai** discusses in detail the challenges of building cyber arms control regimes and formulates solutions, which complement this chapter's recommendations for an inter-Korean cyber arms control agreement. Equally, Michael **Zinkanell**'s chapter on the implementation of cyber arms control in the EU may serve as blueprint for a cyber arms control mechanism on the Korean Peninsula.

For North Korea, such an agreement would give clear legal limitations and reduce its coercive potential against South Korea. Since cyber-enabled capabilities have become North Korea's key asymmetric weapon next to its nuclear arsenal, Pyongyang will need additional security assurances from the US-ROK alliance to agree to give up/ not to use its offensive cyber capabilities. For South Korea, which has been predominantly at the receiving end of cyber attacks, the agreement would provide a framework within which it can cooperate with North Korea to enhance stability on the Peninsula and work towards reconciliation.

The enforcement of the above envisioned cyber arms control regime remains an issue, as political will alone may not suffice to build and maintain a durable agreement. The two Koreas may agree to have their networks and cyber activities monitored by a third party as enforcing organization. A possible candidate could be the United Nations Group of Governmental Experts (UNGGE), which is a UN-mandated working group specializing in developments in the field of information and telecommunications in the context of international security.[38] In section I of her chapter, Cuihong **Cai** introduces various regional and global organizations that could play a role in international cybersecurity cooperation, including the UNGGE. South Korea has been one of the forerunners and important supporters of the UNGGE to strengthen and develop the applicability of international law in cyberspace.

So far, the UNGGE has had difficulties to establish and strengthen international cyberspace law on a global scale. The Korean Peninsula could prove as test bed for cyber arms control agreements on a bilateral level. Since both South and North Korea are members of the UN, the UNGGE could assume jurisdiction over the case and support a control- and enforcement-mechanism for cybersecurity on the Peninsula.

4 Conclusions

The above discussion makes clear that North Korea remains the top security and defence priority for Seoul. North Korea's investments into nuclear-, missile-, and cyber-capabilities in the past years built the foundation of Pyongyang's asymmetry strategy aimed at the US-ROK alliance. It has been shown that the Korean Peninsula is a theatre of heightened cybercrime and cyberterror, predominantly originating from North Korea and targeting South Korea. It is unlikely that North Korea will, in the foreseeable future, give up its nuclear capability completely. If the issue of cybersecurity remains ignored, North Korea is likely to rely even more on its cyber capabilities in the future as this brings many benefits, both politically and financially.

Implementing a cyber arms control agreement on the Korean Peninsula is a difficult task, but it also bears great opportunities and benefits in the long run. The urgency to act is certainly given, as North Korea's continuous cyber attacks against South Korea prevent a peace process and reconciliation. Therefore, it is worth exploring whether a bilateral cyber arms control agreement on the Korean Peninsula could be a building block for establishing confidence between the two Koreas. Such an agreement could minimize the chances of accidental political clashes and military escalation. The case studies of several arms control regimes and current debates to

[38] "UN GGE and OEWG" (Geneva Internet Platform, n.d.), https://dig.watch/processes/un-gge

increase the role of UNGGE in global cybersecurity provide a framework for initial negotiations to take place. The bilateral cyber arms control regime between the two Koreas could, if successful, furthermore become a role model for a multilateral cyber arms control regime, spanning the wider region, or even with a global ambition.

References

Ahn, Seong Kyu. “Cyber Cancer - Can We Prepare for APT (Advanced Persistent Threat).” Column. Seoul: Asan Institute, August 5, 2015.

Bae, Dal Hyung. “North Korea’s Cyber Threat and Options for South Korea’s Policy Improvement and Capability Enhancement.” Seoul, 2011.

CSIS. “Missiles of North Korea.” Missile Defense Project. Washington, DC: Center for Strategic & International Studies, June 14, 2018. https://missilethreat.csis.org/country/dprk/.

Ernst, Maximilian, and Eliana Kim. “Economic Development Under Kim Jong-Un: The Added Value of Traffic Data and Established Indicators in the Study of North Korea’s Economy.” *North Korean Review* 16, no. 2 (2020): 26–48.

Ernst, Maximilian, and Sangho Lee. “Countering Cyber Asymmetry on the Korean Peninsula: South Korea’s Defense Against Cyber Attacks from Authoritarian States.” *Journal for Intelligence, Propaganda and Security Studies* 15, no. 1 (2021): 165–79.

Fraser, Nalani, Jacqueline O’Leary, Vincent Cannon, & Fred Plan. “APT38: Details on New North Korean Regime-Backed Threat Group.” FireEye, October 3, 2018. https://www.fireeye.com/blog/threat-research/2018/10/apt38-details-on-new-north-korean-regime-backed-threat-group.html.

Fuhrmann, Matthew, and Yonatan Lupu. “Do Arms Control Treaties Work? Assessing the Effectiveness of the Nuclear Nonproliferation Treaty.” *International Studies Quarterly* 60, no. 3 (September 2016): 530–39. doi:https://doi.org/10.1093/isq/sqw013.

González, Roberto J. “Hacking the Citizenry?: Personality Profiling, ‘Big Data’ and the Election of Donald Trump.” *Anthropology Today* 33, no. 3 (June 2017): 9–12. doi:https://doi.org/10.1111/1467-8322.12348.

Jun, Jenny, Scott LaFoy, and Ethan Sohn. “North Korea’s Cyber Operations: Strategy and Responses.” Washington, DC: Center for Strategic & International Studies, 2015.

Kim, Jin Moo. “The Reality of North Korea’s Conventional Power: Transition to Asymmetric Strategy.” In *North Korea Conundrum*, edited by Un-Chul Yang, 55–81. Seongnam: Sejong Institute, 2019.

Lee, Sangho. "The Impact of Advanced-Technology Cyber Weapon on the Future Conflict in the Korean Peninsula and East Asia." In *North Korea Conundrum*, edited by Un-Chul Yang, 83–106. Seongnam: Sejong Institute, 2019.

Lee, Sangho. "The Realities of Cyber War and Measures to Counter Future Cyberattacks (in Korean)." *Shidae Jungshin*, Autumn 2009.

Lee, Sangho, Jung-Ho Eom, Bonghan Lee, and Jin-Hyuk Choi. "A Conceptual Study on Knowledge-Based Personal Information Gathering Methods for Conducting Psychological Warfare in Cyberspace." In *Convergence and Hybrid Information Technology*, edited by Geuk Lee, Daniel Howard, Dominik Ślęzak, and You Sik Hong, 287–93. Berlin, Heidelberg: Springer Berlin Heidelberg, 2012. doi:https://doi.org/10.1007/978-3-642-32692-9_37.

Lynn-Jones, Sean M. "A Quiet Success for Arms Control: Preventing Incidents at Sea." *International Security* 9, no. 4 (1985): 154–84.

Mathew, Neil. "$571 Million: Notorious North Korean Hacker Group Has Stolen a Fortune in Cryptocurrency." *CNN*, October 19, 2018. https://www.ccn.com/571-million-notorious-north-korean-hacker-group-has-stolen-a-fortune-in-cryptocurrency/.

Maybaum, Markus, and Jens Tölle. "Arms Control in Cyberspace – Architecture for a Trust-Based Implementation Framework Based on Conventional Arms Control Methods." In *Cyber Power*, 159–73. NATO, 2016.

Meyer, Paul. "Cyber-Security through Arms Control: An Approach to International Co-Opperation." *The RUSI Journal* 156, no. 2 (April 2011): 22–27. doi:https://doi.org/10.1080/03071847.2011.576471.

Palmer, Daniel. "North Korean Hacking Group Lazarus Stole $571 Million in Cryptos: Report." Coindesk, October 19, 2018. https://www.coindesk.com/north-korean-hacking-group-lazarus-stole-571-million-in-cryptos-report.

Richardson, John. "Stuxnet as Cyberwarfare: Applying the Law of War to the Virtual Battlefield." *John Marshall Journal of Computer and Information Law* 29, no. 1 (2011): 1–28.

Rid, Thomas, and Ben Buchanan. "Attributing Cyber Attacks." *Journal of Strategic Attacks* 38 (2015): 1–37.

Rollins, John W. "U.S.-China Cyber Agreement." Federation of American Scientists, October 16, 2015. https://fas.org/sgp/crs/row/IN10376.pdf.

Rosen, Peter. "Competitive Strategies: Theoretical Foundations, Limits, and Extensions." In *Competitive Strategies for the 21st Century: Theory, History, and Practice*, edited by Thomas G. Mahnken, 12–27. Stanford: Stanford University Press, 2012.

Segal, Eddie. "The Impact of AI on Cybersecurity." IEEE Computer Society, n.d. https://www.computer.org/publications/tech-news/trends/the-impact-of-ai-on-cybersecurity.

Shin, Mitch. "What Is the Truth About COVID-19 in North Korea?" *The Diplomat*, January 6, 2021. https://thediplomat.com/2021/01/what-is-the-truth-about-covid-19-in-north-korea/.

Singer, Peter W. "The Cyber Terror Bogeyman." Brookings, November 1, 2012. https://www.brookings.edu/articles/the-cyber-terror-bogeyman/.

Stent, Dylan. "Can the Cyber Terrorism Tactics of North Korea Be Deterred?" *The National Interest*, September 23, 2018. https://nationalinterest.org/blog/skeptics/can-cyber-terrorism-tactics-north-korea-be-deterred-31662.

"UN GGE and OEWG." Geneva Internet Platform, 2019. https://dig.watch/processes/un-gge.

Valeriano, Brandon, and Rayn Maness. *Cyber War versus Cyber Realities: Cyber Conflict in the International System*. Oxford: Oxford University Press, 2015.

"What Did N.Korea's Nuclear Program Cost?" *The Chosun Ilbo*, February 13, 2013. http://english.chosun.com/site/data/html_dir/2013/02/13/2013021301335.html.

Willett, Marcus. "Assessing Cyber Power." *Survival* 61, no. 1 (January 2, 2019): 85–90. doi:https://doi.org/10.1080/00396338.2019.1569895.

Yoo, Seong Un. "Aegis, KF-16 Joint Operations and Core Secrets Were Hacked (in Korean)." *Joongang Ilbo*. September 10, 2015. https://news.joins.com/article/18631228.

Arms Control in the Cyber Domain: A European Approach to Mitigate Digital Threats

Michael Zinkanell

1 Introduction

Humankind is driven by the intrinsic desire to explore and discover. Our creative, analytical, mechanical, and collaborative skills have produced the means and abilities to tame the elements, build megastructures, and to leave our planet for the first time more than 50 years ago. From utilising sticks and stones to programming bits and bytes, all scientific innovations and technical achievements share one decisive common characteristic: human motive gives technology its purpose. The user's intention determines technology's potential, ranging from fruitful opportunities and benefits to harmful threats and dangers. Like a double-edged sword, all technology is in essence simultaneously both constructive and destructive.

M. Zinkanell (✉)
Austrian Institute for European and Security Policy, Wien, Austria
e-mail: michael.zinkanell@aies.at

© The Author(s), under exclusive license to Springer Nature Switzerland AG 2022

G. Boulet et al. (eds.), *Cybersecurity Policy in the EU and South Korea from Consultation to Action*, New Security Challenges,
https://doi.org/10.1007/978-3-031-08384-6_7

Thus, from a simple hammer to nuclear fission, any tool can be transformed into a potential weapon, as history has repeatedly shown.

Our innovative spirit has surpassed the purely analogue realm, opening a new gateway into the digital domain where data and information are created, processed, and communicated predominantly online. As a result, the physical and non-physical worlds have become intertwined, coexisting in an entangled symbiosis. Digitalisation and automation, the modern equivalents of electricity and the steam-engine, represent the cornerstones of the fourth Industrial Revolution. While at the beginning of 2021 more than 4.66 billion users, or 60% of the world population, were connected to the Internet,[1] the number of digital devices connected to the World Wide Web will exceed 75 billion in 2025 (which equals to nine devices per capita).[2] In addition to means of communication such as mobile devices or personal computers, this figure also includes home applications like refrigerators or heating systems and cars. Within the Internet of Things (IoT) many of such non-standard computing devices are already wirelessly connected to one another, exchanging information and data in an automated fashion.

In order to exploit the benefits of technology whilst limiting the risk of any intentional (or accidental) harmful use, political stakeholders and decision makers introduce standards, rules, and regulations. These regulatory measures have often been developed as a response to prior incidents of harmful abuse—various international non-proliferation and disarmament treaties bear witness to this fact. At the same time, restrictive parameters shall not infringe further innovation but create a safe and secure environment in which users can operate and interact with technology effortlessly. For example, the introduction of road safety regulations, including speed limits and regulated vehicle registration, led to fewer traffic fatalities without negatively impacting the automotive industry. Therefore, the same principle should apply to managing the respective threats and dangers of digital information technology. However, without establishing well-defined digital safety measures, including restricting aggressive behaviour,

[1] Simon Kemp, "Digital 2021: Global Overview Report", *Datareportal,* January 27, 2021, https://datareportal.com/reports/digital-2021-global-overview-report

[2] Gilad David Maayan, "The IoT Rundown For 2020: Stats, Risks, and Solutions", January 13, 2020, https://securitytoday.com/articles/2020/01/13/the-iot-rundown-for-2020.aspx

cyberspace will continuously evolve towards a dangerous and hostile environment.

Today, the state-driven weaponisation and militarisation of the cyber domain represent an alarming political and strategic trend that has been accelerating over the last few years.[3] In the most recent past, we have witnessed a steep increase in digital attacks on individuals, enterprises, and states that utilised innovative means to directly inflict harm by sabotaging systems and digital infrastructures, stealing intellectual property or critical information, as well as causing societal destabilisation and disturbance. New ways of unconventional aggressions are carried out by state or state-sponsored actors as well as by organised groups or lone wolves—all of which are acting in a grey area below the threshold of war.[4] This twilight zone is exploited by malicious motivations and hostile actions, which unveil the destructive side of technology ranging from synchronised disruption to subversive threats to national security. From a European perspective, the European Union (EU) and its member states have not only been increasingly targeted by malicious hybrid threats and cyber attacks, the EU is also at the forefront of adopting safety mechanisms as well as rules and regulations to protect European democracy, freedom, and security. Nonetheless, policymakers and stakeholders face an ever-growing difficulty to fully understand the capabilities and implications of emerging destructive technologies due to their fluidity and the accelerating speed of their evolution.

This chapter is arguing that the EU's most prominent cybersecurity strategies, operational instruments, and restrictive measures can be identified as unique arms control mechanisms in the digital realm. In this context, the standards and regulations that the EU has developed on a policy level, which are described by George **Christou** and Ji Soo **Lee**, as well as its counteractive measures are analysed from an arms control perspective. Hereby, the focus will lie on traditional and non-traditional aspects of (cyber) arms control, as well as on the challenges and limits the current measures contain. As the purpose of this chapter goes beyond scrutinising the status quo, a brief analysis of future cyber-related threats of AI and how to address them will be included at the end. The following

[3] CEPS, "Strengthening the EU's Cyber Defence Capabilities, Report of a CEPS Task Force", *Centre for European Policy Studies (CEPS) Brussels* (2018):iv.

[4] Javier Jordan, "International Competition Below the Threshold of War: Toward a Theory of Gray Zone Conflict." *Journal of Strategic Security* 14, no. 1 (2020): 1-24.

subchapter is designed to contextualise cyber threats, providing an insight into the key drivers of digital acceleration and adding a European perspective to the threat environment. The third section, which represents the centrepiece of this chapter, analyses arms control mechanisms within the European cybersecurity strategies and its implementation measures. To conclude, a future outlook into the field of artificial intelligence (AI) calls for further action to counter future AI-related security challenges.

2 Contextualising Cyber Threats

If you know the enemy and know yourself, you need not fear the result of a hundred battles. If you know yourself but not the enemy, for every victory gained you will also suffer a defeat. If you know neither the enemy nor yourself, you will succumb in every battle.[5]

Comprehensively understanding the security threats and challenges posed by the cyber domain, as well as the motives behind attacks, is the first and most crucial step towards defensive and mitigating measures. Thus, in relation to Mason **Richey**'s Cyber Threat Overview in this volume, the nature of these digital threats needs to be put into perspective, with an emphasis on the wider socio-political and economic dependency on digital networks, in order to create European cyber arms control mechanisms.

In the 2010s, cyberspace officially became the fifth domain of warfare, complementing the three traditional domains, land, sea, and air as well as the fourth domain, space.[6] Contrary to the first four domains, cyber is not bound by or subject to geographical or physical borders. Additionally, it is the only sphere that is artificial, hence immaterial and purely human-created. In this sense, the cyber domain does not follow the same laws or physical restrictions that define and limit the nature of all other types of warfare. Consequentially, the vulnerabilities that are targeted and exploited by cyber attacks are, in essence, also virtual. Due to the intertwined human-created symbiosis between the digital and physical world,

[5] Sun Tzu, *The Art of War: Complete Text and Commentaries,* trans. Thomas Cleary (Boston: Shambhala, 2003).

[6] Lesley Seebeck, "Why the fifth domain is different ", *Australian Strategic Policy Institute ASPI*, September 5, 2019, https://www.aspistrategist.org.au/why-the-fifth-domain-is-different/

however, the ramifications of such attacks are affecting the analogue world in various ways.

Two key drivers have enabled this development. First, the digitalisation of information and communication technology (ICT) has fundamentally transformed our social and political lives, while simultaneously representing the backbone of any profitable economy. In both our private and professional lives, we heavily rely on a functioning digital infrastructure to perform the most mundane and complex activities. Today, this transformation has not come to a halt. On the contrary, it is an ongoing process, as digitalisation is further fuelling future innovation: from the latest developments in the fifth generation of telecommunications (5G), representing "one of the most important innovations of our time due to the impact expected in the economy and society"[7] to the ground-breaking field of AI technologies, which "are the most powerful tools in generations for expanding knowledge, increasing prosperity, and enriching the human experience".[8]

Second, rather than enhancing transparency, accuracy, and inclusivity, the growth and speed of digital information and communication have led to "a louder and more chaotic information environment".[9] The massive production, sharing, and processing of digital information has created strong dependencies for our economic and socio-political systems, which in turn must constantly uphold and enlarge the digital infrastructure to safeguard their own viability. Failing to do so, or allowing attackers to disrupt the complex digital chain of events, could lead to a cascading effect with serious consequences on human or national security.

In Europe, Estonia was the first country that was severely affected by a series of cyber attacks in an attempt to destabilise the state in 2007. The attacks, which originated in Russia, targeted multiple institutions such as governmental portals, ministries, Internet providers, banks, media outlets, and private businesses over a span of three weeks and triggered social

[7] ENISA, "ENISA Threat Landscape for 5G Networks - Threat assessment for the fifth generation of mobile telecommunications networks (5G)", European Union Agency for Cybersecurity (ENISA), (November 2019):4.

[8] National Security Commission on Artificial Intelligence, "Final Report", (2021):7 https://www.nscai.gov/wp-content/uploads/2021/03/Full-Report-Digital-1.pdf.

[9] Herbert Lin, "The existential threat from cyber-enabled information warfare", *Bulletin of the Atomic* Scientists", 75(4), (June 2019): 187–196.

unrest and violent rioting.[10] This cyber incident was a wake-up call for the Estonian government and the EU to improve their cybersecurity capacities. For the first time the multifaceted digital vulnerability that derives from cross-sector interdependencies on digital networks became an undeniable reality. This realisation led to a profound change in the security mind-set, comparable to the awareness that motivated decision makers in the past to design conventional arms control mechanisms. It became obvious that the meticulous and synchronised execution of cyber attacks, including its socio-political implications, threatens national security to a degree that requires new measures to regulate the digital domain and control new cyber weapons.

Cyber attacks have become ever more sophisticated and extensive today,[11] while the cyber domain is (ab)used by national actors "as an equaliser for levelling the geopolitical playing field".[12] This dynamic has challenged the EU and its member states to develop and implement concrete strategies and measures to mitigate cyber threats and control the militarisation of the digital environment.

3 European Cybersecurity as Arms Control Mechanisms

Cyber-security includes information and communication security, operational technology and the IT platforms required for digital assets. Cyber-defence is usually understood as including cyber-security, and consisting of threat analyses and strategies to protect against threats directed at citizens, institutions and governments. As for cyber-attacks, these are deliberate activities to disrupt or destroy computer systems and networks. Cyber-deterrence refers to measures for dissuading potential perpetrators through robust systems, sanctions mechanisms and cyber-diplomacy.[13]

[10] NATO Strategic Communications Centre of Excellence, "Hybrid Threats: 2007 cyber-attacks on Estonia", May, 2017.

[11] ENISA, 2020: „ENISA Threat Landscape 2020: Cyber Attacks Becoming More Sophisticated, Targeted, Widespread and Undetected", October, 2020, https://www.enisa.europa.eu/news/enisa-news/enisa-threat-landscape-2020

[12] European Union Institute for Security Studies (EUISS), "Guardian of the Galaxy EU cyber sanctions and norms in cyberspace", *Chaillot Paper* 155, (October 2019):4.

[13] Tania Laţici, "At a Glance – Cyber: How big is the threat?", *European Parliamentary Research Service*, (July 2019):1

One of the primary challenges of approaching cyber-related threats is the lack of a precise definition and the often interchangeably used terms associated with cybersecurity and cyber attacks. The various interpretations of the domain and its terminologies[14] make the immaterial threat even more ambiguous. With regard to cyber arms, one decisive characteristic can be identified in addition to the malicious intention behind the deployment of harmful programmes. Unlike any other type of weapon, cyber attacks are composed of virtual codes, which are weaponised when assembled (programmed) in a certain harmful way.[15] Therefore, traditional measures to limit the production and distribution of weapons, such as arms control, non-proliferation, and disarmament efforts, are not sufficient to constrain the spread of vicious codes in the digital realm.

Against this background, we have to accept a significant challenge for conventional means of comprehending security threats and controlling the creation and distribution of digital weapons. The rapid and interconnected technological changes, which enable the evolution of new possibilities within the digital realm, have already surpassed the means and ends of traditional deterrence and arms control mechanisms. Today's digital security threats are not only wrongfully perceived to be tomorrow's potential risks, but are also met by yesterday's insufficient strategies and approaches. Despite this pessimistic observation, the EU has undertaken profound strategic and operative cybersecurity efforts to improve cyber resilience and cyber defence, which are precisely outlined in Anna-Maria **Osula**'s chapter. This chapter, however, is aiming at identifying these developments as unique approaches to arms control within the digital sphere. Thus, notable aspects of the European cybersecurity strategies and mechanisms will be analysed from an arms control perspective.

The definition of arms control that is used in this context derives mainly from the Organization for Security and Co-operation in Europe (OSCE) and its milestone documents such as the Vienna Document on Confidence- and Security-Building Measures and the Framework for Arms Control (1996). Hereby, specific emphasis is placed on the interplay between

[14] Daniel Schatz; Rabih Bashroush and Julie Wall, "Towards a More Representative Definition of Cyber Security", *Journal of Digital Forensics, Security and Law* 12, no. 2, (2017):8.

[15] Sven Herpig, "Anti-War and the Cyber Triangle. Strategic Implications of Cyber Operations and Cybersecurity for the State", Hull University, January 2016, https://www.stiftung-nv.de/de/publikation/anti-war-and-cyber-triangle-strategic-implications-cyber-operations-and-cyber-security

controlling and limiting the development, construction, distribution, storage, and use of weapons and the creation of confidence- and security-building measures, which are equally relevant. Nonetheless, since hybrid digital threats are fluid and a digital code cannot be compared to a physical apparatus, the concept of arms control is applied in a more flexible fashion. This way the primary measures of European cybersecurity are becoming novel arms control approaches and regulations.

3.1 A European Cyber Arms Control Strategic Framework and Operational Implementation

The EU made its first decisive strategic step towards improving cybersecurity by announcing its cybersecurity strategy in 2013, which bears the title "An Open, Safe and Secure Cyberspace", representing the basis for all future measures. The key acknowledgements that shape the general outline of the 20-page document and provide context to the strategy have remained crucial and valid until today. By outlining the multi-layered and borderless nature of cyberspace as well as stressing the economic reliance on ICT and the new vulnerabilities of digitalisation, the document is taking a broad approach to describing the cybersecurity challenges.[16] Five strategic priorities for preventing and responding to cyber disruptions are outlined: achieving cyber resilience, reducing cybercrime, integrating cyber defence capabilities to the Common Security and Defence Policy, developing cybersecurity resources, and creating a cyberspace policy for the EU promoting its values.[17]

From a novel arms control perspective, the specifics of implementing the strategy are crucial. The key component to strengthen cyber resilience is to develop professional capabilities and enhanced means of cooperation. Therefore, the European Network and Information Systems (NIS) Directive was established in 2016 as the first EU-wide cybersecurity legislation. The Directive calls for national capabilities, cross-border

[16] European Commission, "Joint Communication to the European Parliament, the Council, the European Economic and Social Committee and the Committee of the Regions – Cybersecurity Strategy of the European Union: An Open, Safe and Secure Cyberspace", *JOIN (2013) 1 final*, February 7, 2013.

[17] European Commission, "EU Cybersecurity plan to protect open internet and online freedom and opportunity", *Press Release*, 7 February 2013, https://ec.europa.eu/commission/presscorner/detail/en/IP_13_94.

collaboration, and national oversight of critical sectors.[18] On the national level, member states are required to establish a national NIS strategy, build up cybersecurity capabilities such as computer security incident response services, and organise cyber exercises. Member states shall furthermore collaboratively share information related to cyber risks and incidents as well as best practices on capacity building, training, and raising awareness. National authorities are obliged to supervise the cybersecurity of critical digital sectors and services. This includes the providers of essential services (i.e. water supply, energy, healthcare sector, banking and finance, transport, and all digital infrastructure) as well as digital service providers (i.e. online markets, search engines, cloud computing).[19] This pan-European regulation is the strategic European framework to limit the weaponisation of digital applications. Therefore, it is based on a whole of society approach and includes the governments of member states as well as providers of digital platforms and critical infrastructure.

The European Union Agency for Cybersecurity (ENISA) is the operational arm and institutionalised umbrella authority to the NIS Directive legislation. As the central European institution to assist member states and the private sector to address challenges related to the NIS Directive, ENISA is playing a key role in fostering the implementation of the Directive across the EU. ENISA also facilitated the adoption of the Directive at member states level, thereby translating the strategic cybersecurity framework into national action.

In response to the increasing uncertainty and diminished security,[20] the EU Cybersecurity Act granted ENISA in 2019 a new and extended mandate, increased its remit, and established the first-ever EU-wide framework of cybersecurity certification of ICT products, services, and processes. In order to enable ENISA to run the European cybersecurity certification framework, the technical basis for specific certification schemes had to be

[18] ENISA, "NIS Directive", 2021, https://www.enisa.europa.eu/topics/nis-directive.

[19] Official Journal of the European Union, "Directive (EU) 2016/1148 of the European Parliament and of the Council of 6 July 2016 concerning measures for a high common level of security of network and information systems across the Union", *OJ L 194*, July 19, 2016.

[20] Official Journal of the European Union, "Regulation (EU) 2019/881 of the European Parliament and the Council of 17 April 2019 on ENISA (the European Union Agency for Cybersecurity) and on information and communications technology cybersecurity certification and repealing Regulation (EU) No 526/2013 (Cybersecurity Act)", *L151/15*, (June 2019):4.

created.[21] Each European scheme should set up "the categories of products and services covered; the cybersecurity requirements, such as standards or technical specifications; the type of evaluation, such as self-assessment or third party; and the intended level of assurance".[22]

This development has substantial potential for enhancing EU cyber arms control mechanisms. Based on standards and certifications the composition and application of digital products and services can be regulated. This allows to restrict illicit and potentially harmful activities and programmes. This comes close to an equivalent to traditional arms control and non-proliferation efforts. In the digital non-physical sphere of codes and automated processes, such a holistic cybersecurity certification regime can prevent cyber attacks and cybercrime. In the regulatory process, the EU will engage private entities, international enterprises, and network providers to conform to these new safety precautions. The General Data Protection Regulation (GDPR) can serve as a best practice example, as it turned higher privacy standards and data protection into new digital norms with a global vocation.

The combination of enhancing national cybersecurity capabilities with intensified information sharing paired with an EU-wide cooperation to develop digital standards and regulations safeguards a secure digital ecosystem across the EU. Both the pan-European orientation and the cross-sector approach are innovative characteristics of cyber arms control. The goal of improving knowledge sharing and cooperation, combined with efforts to prevent cyber attacks through regulations, amounts to tangible confidence-building measures, which are a key aspect of traditional arms control measures.

3.2 *Imposing Restrictive Measures as a Cyber Arms Control Mechanism*

The above outlined strategic and operative approaches to developing European cybersecurity measures form innovative arms control mechanisms. Their shortcoming is, however, that they are rather preventive and passive in nature. In order to be prepared to react to or better deter cyber

[21] European Commission, "The EU Cybersecurity Act", 2021, https://digital-strategy.ec.europa.eu/en/policies/cybersecurity-act.

[22] European Commission, "The EU cybersecurity certification framework", 2021, https://digital-strategy.ec.europa.eu/en/policies/cybersecurity-certification-framework.

incidents, the EU has a set of more immediate and proactive means at its disposal, including a typology of malicious actions. They include all unauthorised and illegal activities that access, interfere, or intercept information systems or data.[23] Digital attacks on any critical infrastructure, central ministerial state structures, as well as on essential societal or economic service providers are defined as a threat to member states' security.[24] These strong and frank formulations reflect the growing concern about hostile ambitions and offensive digital capabilities of third states and non-state actors. Therefore, deploying proactive means to directly react to cyber attacks and digital hostilities is an effective way to limit malicious online behaviour.

Building on the EU Cyber Diplomacy Toolbox of 2017[25] as well as on the Council conclusions regarding malicious cyber activities of 2018,[26] the Council of the European Union established a concrete framework in May 2019 that enables "targeted restrictive measures to deter and respond to cyber-attacks".[27] These measures can be taken in reaction to harmful cyber attacks in cases when they are deemed necessary to achieve the objectives of the Common Foreign and Security Policy of the EU.[28] The scope of the new sanctions regime includes all actual and attempted cyber attacks that have a significant impact and, therefore, constitute a serious external threat to the EU and/or its member states—encompassing incidents that "originate or are carried out from outside the EU or use infrastructure outside the EU or are carried out by persons or entities established or operating outside the EU or are carried out with the support of person or entities

[23] Council of the European Union, "Council Decision concerning restrictive measures against cyber-attacks threatening the Union or its Member States", *7299/19*, May 14, 2019.

[24] Council of the European Union, "Council Decision concerning restrictive measures against cyber-attacks threatening the Union or its Member States", *7299/19*, May 14, 2019.

[25] Council of the European Union, "Draft Council Conclusions on a Framework for a Joint EU Diplomatic Response to Malicious Cyber Activities ("Cyber Diplomacy Toolbox") – Adoption", *9916/17*, June 7, 2017.

[26] Council of the European Union, "Council conclusions on malicious cyber activities – approval", *7925/18*, April 16, 2018.

[27] Council of the European Union, "Council Decision concerning restrictive measures against cyber-attacks threatening the Union or its Member States", *7299/19*, (May 2019):4

[28] Council of the European Union, "Council Decision concerning restrictive measures against cyber-attacks threatening the Union or its Member States", *7299/19*, May 14, 2019.

operating outside the EU".[29] Already earlier, the Council of the European Union has adopted the Framework for a Joint EU Diplomatic Response to Malicious Cyber Activities in June 2017. Also known as the Cyber Diplomacy Toolbox, as outlined in Michael **Reiterer**'s chapter on EU cyber diplomacy, the framework is based on the European Commission's presentation of joint response mechanisms to counter cyber operations. By recalling the conclusions of the EU Cybersecurity Strategy (2013) and recognising the intensification of capabilities and ambitions to conduct harmful cyber activities, the document reaffirms the increasing necessity to continue European cybersecurity efforts.[30]

The main purpose of the Toolbox is to provide a framework that enables reacting to cyber attacks and malicious operations diplomatically within the scope of the European Common Foreign and Security Policy. The deterring actions range from diplomatic statements of condemnation to more restrictive measures such as imposing sanctions.[31] Sanctions can be applied to individuals and institutions held accountable for cyber attacks. The purview of sanctionable actions includes any behaviour that is facilitating or sustaining digital attacks financially, technically, or physically. In 2022, the EU decided to prolong the effective framework of restrictive measures against cyber threats, which currently apply to eight individuals and four entities, for a further three years.[32]

Although the number of sanctioned entities seems small measured against the multitude of cyber attacks on the EU and its member states, sanctions are limiting the amount and intensity of digital hostility due to their deterring effect on potential aggressors. The various possibilities of diplomatic condemnation and restrictive counteractions are, therefore, a traditional but still effective way of controlling the creation and

[29] Council of the European Union, "Cyber-attacks: Council is now able to impose sanctions", Press release, May 17, 2019, https://www.consilium.europa.eu/en/press/press-releases/2019/05/17/cyber-attacks-council-is-now-able-to-impose-sanctions/.

[30] Council of the European Union, "Draft Council Conclusions on a Framework for a Joint EU Diplomatic Response to Malicious Cyber Activities ("Cyber Diplomacy Toolbox") – Adoption", *9916/17*, June 7, 2017.

[31] Council of the European Union, "Draft Council Conclusions on a Framework for a Joint EU Diplomatic Response to Malicious Cyber Activities ("Cyber Diplomacy Toolbox") – Adoption", *9916/17*, Brussels, June 7, 2017.

[32] Council of the European Union, "Cyber-attacks: Council extends sanctions regime until 18 May 2025", Press Release, Brussels, 16 May, 2022, https://www.consilium.europa.eu/en/press/press-releases/2022/05/16/cyber-attacks-council-extends-sanctions-regime-until-18-may-2025/.

deployment of cyber weapons. As attribution of cyber attacks is the problematic hurdle to justify countermeasures, the Cyber Diplomacy Toolbox has been skilfully crafted to allow the implementation of measures without explicit attribution. Due to their intangible nature, it is exceptionally difficult to attribute malicious actions in cyberspace accurately to their source of origin. Even in cases when the aggressor can be identified, "the evidence on which the actual attribution is made is not generally public information".[33] This complicates any responsive diplomatic measures. Therefore, the EU clarifies that the act of attribution is a sovereign decision; consequently, "not all measures of a joint EU diplomatic response to malicious cyber activities require attribution to a State or a non-State actor".[34]

4 Conclusion and Future Outlook into the Field of Artificial Intelligence

In conclusion, the EU has developed strategic frameworks, operative mechanisms, and restrictive instruments to successfully mitigate cyber threats in a preventive and proactive fashion, resembling aspects of arms control and non-proliferation. The focus is put on two aspects: the limitation of potentially harmful programmes by enforcing standards and regulations as well as on restricting the malicious behaviour and intent. The latter overcomes the issue of fluidity and ambiguity within the digital realm by emphasising the human element: technological features become less relevant, since individual human action shapes and defines them. Thus, establishing parameters of digital behaviour and rules of cyber engagement is the decisive key to cyber arms control, as Sangho **Lee** and Maximilian **Ernst** show in discussing the case of the Korean Peninsula. However, as automation, machine learning and the entire sphere of AI are rapidly growing, a completely new dimension of non-human reasoning and computerised decision-making has to be considered in current and future cybersecurity and cyber arms control.

[33] Katriina Härmä and Tomáš Minárik,"European Union Equipping Itself against Cyber Attacks with the Help of Cyber Diplomacy Toolbox", The NATO Cooperative Cyber Defence Centre of Excellence, June 19, 2017, https://ccdcoe.org/incyder-articles/european-union-equipping-itself-against-cyber-attacks-with-the-help-of-cyber-diplomacy-toolbox/.

[34] Council of the European Union, "Draft Council Conclusions on a Framework for a Joint EU Diplomatic Response to Malicious Cyber Activities ("Cyber Diplomacy Toolbox") – Adoption", *9916/17*, (June 2017):5.

Artificial Intelligence is clearly on the verge of revolutionising digitalisation and the nature of information and communication technology. We are beginning to experience the full potential of data, which is opening a wide range of futuristic scenarios and their socio-political and security implications. Global power competition is progressively centring around the means and abilities to improve AI capabilities—also in the field of Autonomous Weapons Systems (AWS). Arguing from the perspective of cybersecurity and cyber arms control, it is safe to say that through this race for technological superiority new defensive and offensive capabilities are emerging. This leads to one clear and crucial deduction for the EU: it is the responsibility of today's political decision makers to agree on the means and ends of using AI in the future. If the range of applications that AI offers remains unregulated, the harmful abuse and exploitation of its capabilities become inevitable. In order to prevent the risks and threats of AI, it is necessary to scrutinise the intersection of AI, cybersecurity, and cyber arms control mechanisms.

The EU has already started to deal strategically with the multifaceted field of AI. The most prominent developments include the European Commission's *White Paper on Artificial Intelligence* and the *Report on the Safety and Liability Implications of Artificial Intelligence, the Internet of Things and Robotics*, both published in 2020, and the release of the Commission's AI package in April 2021. The latter includes three key documents: a Communication on Fostering a European Approach to AI, the updated version of the Coordinated Plan with member states, and a Proposal for an AI Regulation. Without going into the details of all these documents, as that would go beyond the scope of this contribution, three central elements shall be highlighted.

First, safeguarding cybersecurity measures is vital to ensuring the resilience and operability of AI systems.[35] Cyber attacks targeting and manipulating AI applications could lead to wide-ranging destabilisation, altering the functionality and effectiveness of AI tremendously. All AI systems shall thus be regarded as extremely vulnerable to cyber attacks, requiring the implementation of necessary state-of-the-art cybersecurity mechanisms to all AI and their underlying ICT infrastructure.

[35] European Commission, "Proposal for a Regulation of the European Parliament and of the Council Laying Down Harmonised Rules on Artificial Intelligence (Artificial Intelligence Act) and Amending Certain Union Legislative Acts", (2021): 30.

Second, as automation is a defining characteristic and key driver of any AI—its safety relies on foreseeing and determining potential risks of unintended harmful outcomes in advance. In particular within the context of military systems and automated weapons, controllability and predictability must be assured at any time, requiring human control and compliance with international humanitarian law.[36] Thus, the principle of autonomy must correlate with the ability of responsible and effective human oversight, including a detailed risk assessment for all AI systems.

Third, since the current geopolitical tensions render a global regulation of AI very unlikely, the EU should take the lead in introducing AI standards and safety mechanisms, which could serve as role models. Against this background, reassuring the compliance with the current framework of International Humanitarian Law is absolutely necessary for all military applications of AI, especially autonomous weapons systems. Thereby, certain criteria such as the ability to predict AI behaviour and actions, assuring human control over operational decisions, and the necessity to trace AI performances and effects to the responsible human agent for accountability reasons need to be met and ensured.

References

Centre for European Policy Studies (CEPS), "Strengthening the EU's Cyber Defence Capabilities, Report of a CEPS Task Force", November 2018, Centre for European Policy Studies (CEPS) Brussels.

Council of the European Union, "Cyber-attacks: Council is now able to impose sanctions", *Press release*, 17 May 2019a. https://www.consilium.europa.eu/en/press/press-releases/2019/05/17/cyber-attacks-council-is-now-able-to-impose-sanctions/.

Council of the European Union, "Cyber-attacks: Council extends sanctions regime until 18 May 2025", Press Release, Brussels, 16 May, 2022, https://www.consilium.europa.eu/en/press/press-releases/2022/05/16/cyber-attacks-council-extends-sanctions-regime-until-18-may-2025/.

Council of the European Union, "Draft Council Conclusions on a Framework for a Joint EU Diplomatic Response to Malicious Cyber Activities ("Cyber Diplomacy Toolbox") – Adoption", *9916/17*, Brussels, June 7, 2017a.

[36] Martin Hagström, "Military applications of machine learning and autonomous systems", in *The Impact of Artificial Intelligence on Strategic Stability and Nuclear Risk, Volume I Euro-Atlantic Perspectives*, ed. Vincent Boulanin, May 2019.

Council of the European Union. "Council conclusions on malicious cyber activities – approval", *7925/18*, Brussels, April 16, 2018.
Council of the European Union. "Council Decision concerning restrictive measures against cyber-attacks threatening the Union or its Member States", *7299/19*, Brussels, May 14, 2019b.
Council of the European Union. "Draft Council Conclusions on a Framework for a Joint EU Diplomatic Response to Malicious Cyber Activities ("Cyber Diplomacy Toolbox") – Adoption", *9916/17*, Brussels, June 7, 2017b.
European Union Agency for Cybersecurity (ENISA). "ENISA Threat Landscape for 5G Networks - Threat assessment for the fifth generation of mobile telecommunications networks (5G)", *European Union Agency for Cybersecurity (ENISA)*, November 2019, p. 4.
European Commission, "EU Cybersecurity Plan to Protect Open Internet and Online Freedom and Opportunity." *European Commission Press Corner*, February 7, 2013a. European Commission. https://ec.europa.eu/commission/presscorner/detail/en/IP_13_94.
European Commission. "Joint Communication to the European Parliament, the Council, the European Economic and Social Committee and the Committee of the Regions – Cybersecurity Strategy of the European Union: An Open, Safe and Secure Cyberspace", *JOIN (2013) 1 final*, Brussels, (February 7, 2013b). https://eur-lex.europa.eu/legal-content/EN/TXT/?uri=celex%3A52013JC0001.
European Commission. "Proposal for a Regulation of the European Parliament and of the Council Laying Down Harmonised Rules on Artificial Intelligence (Artificial Intelligence Act) and Amending Certain Union Legislative Acts", *COM 206 final*, Brussels, April 21, 2021a.
European Commission. "The EU Cybersecurity Act", 2021b. https://digital-strategy.ec.europa.eu/en/policies/cybersecurity-act.
European Commission. "The EU cybersecurity certification framework", 2021c. https://digital-strategy.ec.europa.eu/en/policies/cybersecurity-certification-framework.
European Union Agency for Cybersecurity (ENISA). "ENISA Threat Landscape 2020: Cyber Attacks Becoming More Sophisticated, Targeted, Widespread and Undetected", (October 2020), https://www.enisa.europa.eu/news/enisa-news/enisa-threat-landscape-2020.
European Union Agency for Cybersecurity (ENISA). "NIS Directive", 2021. https://www.enisa.europa.eu/topics/nis-directive.
European Union Institute for Security Studies (EUISS). "Guardian of the Galaxy EU cyber sanctions and norms in cyberspace." *Chaillot Paper* 155 (October 2019).
Hagström, Martin. "Military applications of machine learning and autonomous systems". In *The Impact of Artificial Intelligence on Strategic Stability and*

Nuclear Risk, edited by Vincent Boulanin, Volume I Euro-Atlantic Perspectives, *Stockholm International Peace Research Institute*, May 2019.

Härmä, Katriina, and Tomáš Minárik. "European Union Equipping Itself against Cyber Attacks with the Help of Cyber Diplomacy Toolbox", *The NATO Cooperative Cyber Defence Centre of Excellence*, https://ccdcoe.org/incyder-articles/european-union-equipping-itself-against-cyber-attacks-with-the-help-of-cyber-diplomacy-toolbox/.

Herpig, Sven. "Anti-War and the Cyber Triangle. Strategic Implications of Cyber Operations and Cybersecurity for the State", Hull University (January 2016). https://www.stiftung-nv.de/de/publikation/anti-war-and-cyber-triangle-strategic-implications-cyber-operations-and-cyber-security.

Jordan, Javier. "International Competition Below the Threshold of War: Toward a Theory of Gray Zone Conflict." *Journal of Strategic Security* 14, no. 1 (2020): 1-24.

Kemp, Simon, "Digital 2021: Global Overview Report", *DATAREPORTAL*, 27 January, 2021. https://datareportal.com/reports/digital-2021-global-overview-report.

Lațici, Tania. "At a Glance – Cyber: How big is the threat?" *EPRS | European Parliamentary Research Service*, (July 2019).

Lin, Herbert. "The existential threat from cyber-enabled information warfare." *Bulletin of the Atomic Scientists* 75, no. 4, (2019): 187–196.

Maayan, Gilad David, "The IoT Rundown For 2020: Stats, Risks, and Solutions", *Security Today*, 2020. https://securitytoday.com/articles/2020/01/13/the-iot-rundown-for-2020.aspx

National Security Commission on Artificial Intelligence, "Final Report" (2021): p. 7. https://www.nscai.gov/wp-content/uploads/2021/03/Full-Report-Digital-1.pdf.

NATO Strategic Communications Centre of Excellence, 2019: "Hybrid Threats: 2007 cyber attacks on Estonia" in Aday, Sean, and Māris Andžāns, Una Aleksandra Bērziņa Čerenkova, Francesca Granelli, John-Paul Gravelines, Mils Hills, Miranda Holmstrom, Adam Klus, Irene Martinez-Sanchez, Mariita Mattiisen, Holger Molder, Yeganeh Morakabati, James Pamment, Aurel Sari, Vladimir Sazonov, Gregory Simons, andJonathen Terra, "Hybrid Threats. A Strategic Communications Perspective", (2019).

Official Journal of the European Union. "Directive (EU) 2016/1148 of the European Parliament and of the Council of 6 July 2016 concerning measures for a high common level of security of network and information systems across the Union", *OJ L 194*, (July 19, 2016). https://eur-lex.europa.eu/eli/dir/2016/1148/oj.

Official Journal of the European Union. "Regulation (EU) 2019/881 of the European Parliament and the Council of 17 April 2019 on ENISA (the European Union Agency for Cybersecurity) and on information and communications technology cybersecurity certification and repealing Regulation (EU) No 526/2013 (Cybersecurity Act)", *L151/15*, (June 7, 2019). https://eur-lex.europa.eu/legal-content/EN/TXT/?uri=uriserv:OJ.L_.2019.151.01.0015.01.ENG.

Schatz, Daniel, Rabih Bashroush, and Julie Wall. "Towards a More Representative Definition of Cyber Security", *Journal of Digital Forensics, Security and Law* 12, no. 2, Article 8, (2017).

Seebeck, Lesley. "Why the fifth domain is different." Australian Strategic Policy Institute ASPI, (2019). https://www.aspistrategist.org.au/why-the-fifth-domain-is-different/.

Sun Tzu, *The Art of War: Complete Text and Commentaries*, trans. Thomas Cleary (Shambhala, 2003).

South Korea's Cyber Defence and Digital Cooperation with the EU

Kyu-dok Hong and Seong-jong Song

1 Introduction

Once derided a "laughing stock" by the US intelligence community, North Korea's cyber power has become a force to be reckoned with.[1] In its report of September 2020, Microsoft pointed out a significant spike in hacking attempts to public policy groups and organizations involved in

[1] David E. Sanger et al., "The World Once Laughed at North Korean Cyberpower. No More," *New York Times*, October 15, 2017, https://www.nytimes.com/2017/10/15/world/asia/north-korea-hacking-cyber-sony.html.

K.-d. Hong (✉)
Sookmyung Women's University, Seoul, South Korea
e-mail: kdhong@sookmyung.ac.kr

S.-j. Song (✉)
Daejeon University, Daejeon, South Korea

© The Author(s), under exclusive license to Springer Nature Switzerland AG 2022

G. Boulet et al. (eds.), *Cybersecurity Policy in the EU and South Korea from Consultation to Action*, New Security Challenges,
https://doi.org/10.1007/978-3-031-08384-6_8

covid-19 research.[2] It also revealed that more than half of all nation-state targeting originated in Russia, with the other half coming mostly from North Korea, China, and Iran.[3] In a similar vein, the US Government Accountability Office (GAO) report also released in December 2020 warned that "supply chains are being targeted by increasingly sophisticated threat actors, including foreign cyber threat nations such as Russia, China, Iran and DPRK."[4]

The GAO report provides the first clues to a baffling question revolving around the recent cyber attack from Russia: How have the Russians successfully hacked into the computer networks of so many US government agencies and not been detected for so many months? Hackers ostensibly linked to Russia's SVR (Foreign Intelligence Service) employed a novel ***modus operandi*** of targeting vulnerabilities in supply chain by using a malicious code that circumvented the US government's multibillion-dollar Einstein system which heavily relies on detecting known viruses.[5] This incident might be interpreted as an anecdotal evidence of Cold War 2.0 in the cyberspace between global powers. Main objectives of a Cyber Cold War are intelligence gathering, intellectual property theft, physical sabotage, and influence operations.[6] The Biden administration's project of hosting a global Summit for Democracy "to strengthen our democratic institutions, honestly confront the challenge of nations that are backsliding, and forge a common agenda to address threats to our common values"[7] could be well placed to take up cyber threats stemming from adversaries, including Russia, China, North Korea, and Iran.[8]

[2] Microsoft, *Microsoft Digital Defense Report*, September 2020, https://www.microsoft.com/en-us/download/confirmation.aspx?id=101738.

[3] Maggie Miller, "Microsoft Reports Spike in Foreign Targeting of COVID-19 Researchers, Think Tanks," *Hill*, September 29, 2020.

[4] U.S. Government Accountability Office (GAO), "Information Technology: Federal Agencies Need to Take Urgent Action to Manage Supply Chain Risks," *GAO-21-171*, December 2020, https://www.gao.gov/assets/720/711266.pdf.

[5] David E. Sanger et al., "Billions Spent on U.S. Defenses Failed to Detect Giant Russian Hack," *New York Times*, December 16, 2020. https://www.nytimes.com/2020/12/16/us/politics/russia-hack-putin-trump-biden.html.

[6] Bertrand Venard, "The Cold War 2.0 between China and the US Is Already a Virtual Reality," *Conversation*, October 17, 2019.

[7] Biden, Joe, "The Power of America's Example: The Biden Plan for Leading the Democratic World to Meet the Challenges of The 21St Century," 2020, https://joebiden.com/americanleadership/.

[8] For the cyber threat overview in 2021, see Mason **Richey** in this volume.

The former British Prime Minister Boris Johnson proposed expanding the G7 (more than half of them belong to EU) to a D-10, by including South Korea, India, and Australia.[9] Such a grouping could be an umbrella mechanism of the Summit for Democracy.

Recognizing the importance of cybersecurity in the overall context of security cooperation, the EU and South Korea have embarked on a pilot project on cybersecurity in implementing the EU's security cooperation project based on the 2018 policy paper, "Enhancing Security cooperation in and with Asia." Korea is a co-genial partner with the highest rates of Internet penetration and smartphone adoption in the world.[10] Furthermore, in 2020 South Korea published its first comprehensive *National Cybersecurity Strategy*,[11] while the EU unveiled its *Cybersecurity Strategy for the Digital Decade*.[12] Both documents highlight the importance of international cooperation to bolster global cybersecurity and indicate a growing cyber threat from North Korea which operates about 7000 cyberwarriors capable of performing a wide range of illegal activities including theft, denial of service (DDoS), espionage, and sabotage.[13]

In that sense, this chapter is to examine the status of cyber threats stemming from North Korea on the one hand, and to explore ways to strengthen cyber defence between South Korea and the EU.

After an introduction, the evolving North Korean cyber threats are demonstrated, focusing on the role of the Reconnaissance General Bureau (RGB) and its cyber footprints. The focus on the North Korean threat factor is based on the facts that South Korea is North Korea's primary target and that it is easier for North Korea to attack South Korea as one of the most wired countries in the world. The third section examines the possibility of an EU-South Korea cyber-strategic partnership based on the respective cybersecurity strategies. To this end, the conclusions formulate some policy recommendations to improve South Korea's cybersecurity defence posture and its relevance for further cyber cooperation between South Korea and the EU.

[9] Ishaan Tharoor, "An Emerging New Alliance of Democracies," *Washington Post*, December 18, 2020.

[10] Annie Palmer, "Coupang, a SoftBank-backed start-up, is crushing Amazon to become South Korea's biggest online retailer," *CNBC*, June 16, 2020.

[11] See Anna-Maria **Osula** in this volume.

[12] On the EU strategy, see Michael **Reiterer** in this volume.

[13] Chong-woo Kim and Carolina Polito, "The Evolution of North Korean Cyber Threats," *Asan Institute for Policy Studies*, 20 February 2019, p. 1. http://en.asaninst.org/contents/the-evolution-of-north-korean-cyber-threats/.

2 Evolving North Korea's Cyber Threats

On April 15, 2020, the US Departments of State, Homeland Security, and Treasury, and the Federal Bureau of Investigation issued a joint advisory to raise awareness of the seriousness of the cyber threat posed by North Korea. It highlighted North Korea's cyber activities around the world, identified US government resources that provide technical and threat information, and included measures to counter the cyber threat. More specifically, the CISA (Cybersecurity and Infrastructure Security Agency) warned that North Korea poses "a significant threat to the integrity and stability of the international financial system." It further noted that "under the pressure of robust U.S. and UN sanctions, the DPRK has increasingly relied on illicit activities—including cybercrime—to generate revenue for its weapons of mass destruction and ballistic missile programs."[14]

An advisory of CISA included five incidents in which North Korea was publicly attributed as "state-sponsored cyber actors and co-conspirators":

1. Sony Pictures. In November 2014, North Korea's cyber actors allegedly launched a cyber attack on Sony Pictures in retaliation for the 2014 film *The Interview*.
2. Bangladesh Bank Heist. In February 2016, North Korea's cyber actors allegedly attempted to steal at least $1 billion from financial institutions across the world and allegedly stole $81 million from the Bangladesh Bank.
3. WannaCry 2.0. North Korea's cyber actors developed the ransomware known as WannaCry 2.0, as well as two prior versions of the ransomware, which infected hundreds of thousands of computers in hospitals, schools, businesses, and homes in over 150 countries.
4. FASTCash Campaign. Since late 2016, North Korean state-sponsored cyber actors have employed a fraudulent ATM cash withdrawal scheme known as "FASTCash" to steal tens of millions of dollars from ATMs in Asia and Africa.

[14] CISA (*Cybersecurity and Infrastructure Security Agency*), "Alert (AA20-106A): Guidance on the North Korean Cyber Threat," April 15, 2020, https://us-cert.cisa.gov/ncas/alerts/aa20-106a.

5. Digital Currency Exchange Hack. In April 2018, North Korea's cyber actors hacked into a digital currency exchange and stole nearly $250 million worth of digital currency. In March 2020, the Department of the Treasury announced that 113 digital currency accounts owned by individuals of North Korea who had been previously indicted on money laundering and unlicensed money transmitting charges were subject to forfeiture.[15]

The inter-agency report was prepared before the outbreak of covid-19. The report stopped short of accusing North Korea of expanding its army of hackers to take advantage of the ongoing global crisis. But it specifically pointed out that state-sponsored North Korean hackers had "demonstrated a pattern of disruptive and harmful cyber activity that is wholly inconsistent with the growing international consensus on what constitutes responsible state behavior in cyberspace."[16]

The few cases mentioned above for illustration are just the tip of the iceberg. Nevertheless, cyber attacks are in most cases difficult to pin on Pyongyang because they can be launched also in third countries like China and Russia. From North Korea's standpoint, cyber attacks are in fact levelling the playing field for them vis-à-vis the US, South Korea, and other developed countries in a way that nuclear weapons cannot: unlike nuclear weapons, cyber "weapons" are not an all-or-nothing proposition.[17] Ironically, South Korea is easier to attack as it is one of the most wired countries in the world: This high level of connectivity is a double-edged sword: while contributing to the efficiency of the economy and the development of the Fourth Industrial Revolution, it renders South Korea also vulnerable because of the ubiquity of cyber-tools.

2.1 From a Threatening Weapon to an All-Purpose Sword

ICT is the dual use good par excellence: While it is an essential means for technological and thereby economic development and growth, it can also be used by authoritarian regimes as a means of oppression and control. If

[15] Ibid.

[16] David E. Sanger and Nicole Perlroth, "U.S. Accuses North Korea of Cyberattacks, a Sign That Deterrence Is Failing," *New York Times*, April 15, 2020.

[17] Ellen Ioanes, "Kim Jong Un has quietly built a 7,000-man cyber army that gives North Korea an edge nuclear weapons don't," *Business Insider*, July 17, 2020.

the latter fails and information starts to flow countering propaganda, this can destabilize governments.[18] For this reason access to the Internet in North Korea was strictly limited even for the elites to maintain control and secure the survival of the regime until 2010.[19] After Kim Jong-un was designated as heir-apparent to former leader Kim Jong-il, North Korea's Internet policy showed a marked difference from that of his predecessor. On the occasion of the 65th anniversary of the founding of the Workers' Party of Korea in October 2010, some 80 foreign journalists were invited to cover the military parade. During these celebrations, "DPRK is said to have made its first full connection to the Worldwide Web from its own territory."[20]

The 1991 Gulf War was a watershed leading to "a sea change" in North Korea's perception to cyber warfare. Kim Jong-il, then North Korea's leader, realized the advantage of having a networked military force and ordered to build up cyber capabilities. Another lesson was provided by the Iraq War. Kim Jong-il stated in 2003, "If the Internet is like a gun, cyber-attacks are like atomic bombs"; and "a war in the 21st century will be decided by one's conduct of electronic warfare," thus "cyber units should be my detached force and backup power."[21]

Kim Jong-un, the current leader, followed this path and built up mission-oriented cyber units. The South Korean National Intelligence Service quoted Kim Jong-un as saying in 2013 that "Cyberwarfare, along with nuclear weapons and missiles, is an 'all-purpose sword' that guarantees our military's capability to strike relentlessly."[22]

Therefore, cyber operations are an extension of North Korea's broader national strategy. During peacetime, cyber capabilities allow North Korea to upset the status quo with little risk of retaliation or immediate operational risk. During wartime, North Korea could target US and South

[18] Kyungmin Ko and others, "The Internet dilemma and control policy: political and economic implications of the Internet in North Korea," *Korean Journal of Defense Analysis*, Vol. 21, No. 3 (September 2009):279–280.

[19] Martyn Williams, "How the Internet Works in North Korea," *Slate*, November 28, 2016. https://slate.com/technology/2016/11/how-the-internet-works-in-north-korea.html.

[20] Reporters Without Borders, "Internet Enemies 2011 – North Korea," March 11, 2011, https://www.refworld.org/docid/4d82268ec.html.

[21] Kong, Ji-young et al., "The All-Purpose Sword: North Korea's Cyber Operations and Strategies," *11th International Conference on Cyber Conflict (CyCon)*, Tallinn, Estonia, 2019, https://ccdcoe.org/uploads/2019/06/Art_08_The-All-Purpose-Sword.pdf.

[22] Sanger at al, "World."

Korean command, control, communications, computers, intelligence, surveillance, and reconnaissance (C4ISR) in support of its "quick war, quick end" (속전속결) strategy. North Korea's cyber doctrine, if one exists, may be premised on the concept that a networked or net-centric military like that of South Korea is vulnerable to cyber capabilities.[23] Such vulnerability begs for revamping existing South Korea-US cooperation networks and establishing a new cyber cooperation command centre within the South Korea-US Combined Forces Command.

2.2 A Role of the Reconnaissance General Bureau (RGB)

North Korea's military, even with its obsolete fighter planes, warships and tanks, and malnourished soldiers, has become a force to be reckoned with through persistent nurturing more than 7000 "cyberwarriors," most of whom are under the control of the Reconnaissance General Bureau (RGB, 정찰총국). The RGB is North Korea's primary organ in charge of intelligence and cyber operations as well as conventional terrorist and clandestine activity, which are carried out outside North Korea. Historically, it has been linked to numerous peacetime commando raids, infiltrations, disruptions, and other secretive operations, such as the sinking of South Korea's *Cheonan* naval vessel in 2010 by torpedo-ambush, and more historically, the terrorist-style bombings in Rangoon 1983 and the Blue House Raid to assassinate the Korean President in 1968.[24] The RGB controls the bulk of known North Korean cyber capabilities, mainly under Bureau 121.[25]

As it is the case with all North Korea institutions, the organization of the RGB remains opaque. After a major restructuring, the RGB currently commands six departments, so-called departments of Operations, Reconnaissance, Foreign Intelligence, Inter-Korean Affairs, Technical, and Logistics & Support Services, respectively. The most principal cyber unit is Bureau 121, which is also called "Electronic Reconnaissance

[23] Jun et al., *North Korea's Cyber Operations: Strategy and Response* (New York: CSIS, 2015), http://csis-website-prod.s3.amazonaws.com/s3fs-public/legacy_files/files/publication/151216_Cha_NorthKoreasCyberOperations_Web.pdf.

[24] Jun et al, "The Organization of Cyber Operations in North Korea," *CSIS (Korea Chair Platform)*, December 18, 2014, https://csis-website-prod.s3.amazonaws.com/s3fs-public/legacy_files/files/publication/141218_Cyber_Operations_North_Korea.pdf.

[25] Jun et al., *North Korea's Cyber Operations: Strategy and Response* (New York: CSIS, 2015), http://csis-website-prod.s3.amazonaws.com/s3fs-public/legacy_files/files/publication/151216_Cha_NorthKoreasCyberOperations_Web.pdf.

Bureau's Cyber Warfare Guidance Bureau" [사이버전지도국]. The RGB probably consisted of seven bureaus and is known to have added a 7th (Bureau 121) sometime after 2013. Bureau 121 is probably North Korea's primary cyber operations apparatus.[26] First, Unit 91 which came into being through a cyber revamp during 2014~2015, was initially tasked to perform cyber espionage operations, aiming at government, corporate, and public targets of South Korea. However, from 2014, its focus has shifted to "acquiring advanced technologies needed for nuclear development and long-range missiles from developed countries."[27] Second, Unit 31 is responsible for malware development, while Unit 32 is responsible for software development for military use.[28] Third, the Investigation Office (자료조사실) is in charge of extensive hacking activities, especially targeting the US Congress or South Korea's national assembly, political parties, economic think tanks, and other academic institutions of the US and South Korea.[29] Fourth, Lab 110, the main cyber unit under the RGB, applies cyber attack techniques to conduct intelligence operations.[30] This unit is believed to possess hacking techniques sophisticated enough to conduct destructive attacks on South Korea's financial institutions.[31] Fifth, Unit 180, the main overseas intelligence agency of the RGB, is known to have engaged in hacking financial institutions by breaching and withdrawing money out of bank accounts. It earned its notorious reputation in the $81 million cyber heist at the Bangladesh bank and the 2014 attack on Sony's Hollywood studio (Fig. 1).[32]

In general, North Korea's cyber operations reflect at least three distinct characteristics. First, North Korea's hacking groups and cyber units have demonstrated significant diversity in terms of their expertise and capabilities, which makes attribution even more onerous and tricky. The boundary

[26] Recorded Future, "North Korea Cyber Activity," *Recorded Future Insikt Group Report*, June 14, 2020, pp. 6–7. https://go.recordedfuture.com/hubfs/reports/north-korea-activity.pdf.

[27] Steve Miller, "Where Did DPRK's Cyber Army Come From?" *VOA News*, November 20, 2018.

[28] Jun et al., "*North Korea's*", p. 47.

[29] Young-jong Lee, "Cyber warfare is a merciless treasure-sword of the Korean People's Army," *Sisa Journal*, December 30, 2014, http://www.sisajournal.com/news/articleView.html?idxno=140623.

[30] Kong et al., "All-purpose," p. 5.

[31] Jun et al., "*North Korea's*", p. 42.

[32] Ju-min Park and James Pearson, "Exclusive: North Korea's Unit 180, the cyber warfare cell that worries the West," *Reuters*, May 21, 2017.

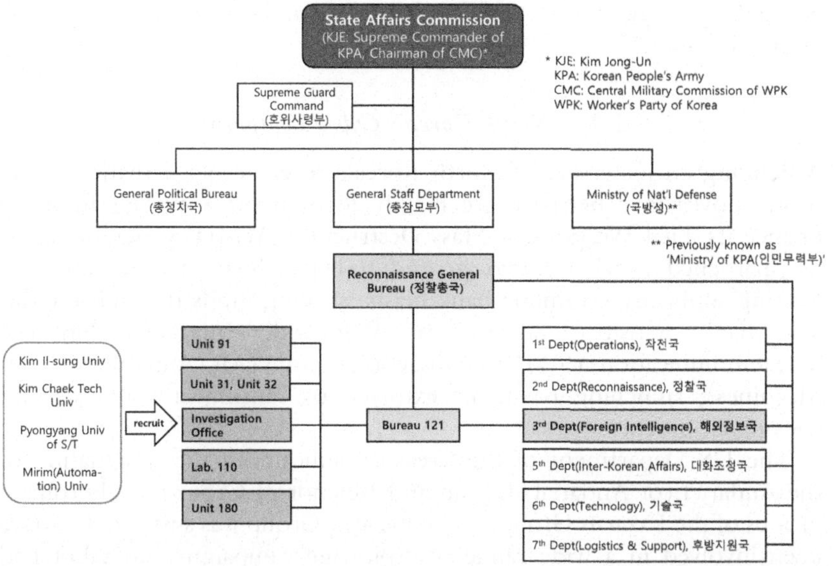

Fig. 1 Organization chart of the RGB. (Source: Adapted from various sources. Mathew Ha and David Maxwell, Kim Jong-un's "All-Purpose Sword" (Washington DC: Foundation for Defense of Democracies, 2018), p. 12, https://www.fdd.org/wp-content/uploads/2018/09/REPORT_NorthKorea_CEEW.pdf)

between high-end and low-end cyber operations has usually been obscured. North Korea recruited non-state actors and entities as surrogate to provide cover for their operators. It also utilizes low-cost/high-reward, and off-the-shelf techniques such as distributed denial-of-service (DDoS) attacks that are freely available. Second, North Korea has increasingly divulged a resolve for cyber escalation by means of targeting the critical infrastructure of other nations, along with private corporations and financial institutions. Lately, North Korea has shown a great interest in acquiring illicit financial gain by circumventing multi-layered international sanction regimes and generating foreign currency. Third, from North Korea's standpoint, cyberspace is a "lopsided playing field" to its advantage. North Korea's Internet infrastructure is almost completely insulated from global networks because its entire Internet traffic goes through only two providers: China's Unicom and Russia's TransTeleCom. The country is largely unplugged from the global Internet and is protected by China's "Great Firewall."[33]

2.3 North Korea's Cyber Footprints

A defining characteristic of North Korea's recent cyber activities is that those moves in cyberspace are mainly focused on acquiring funds to finance its illicit Weapons of Mass Destruction (WMD) programmes. It has sharpened its cyber ***modus operandi*** to bypass financial sanctions, conducting sophisticated online bank heists; stealing funds through fraudulent bank transfers, Society for Worldwide Interbank Financial Telecommunications (SWIFT) transactions, and ATMs (Automated Teller Machines); launching payments extorted by ransomware attacks; and hacking cryptocurrencies.[34]

The US Department of the Treasury announced the designation of the Andariel (or Andarial) Group and Bluenoroff Group as subordinate groups of the Lazarus Group.[35] The Lazarus Group was suspected to have been involved in a wide range of espionage campaigns and sabotaged operations of financial institutions, media stations, and manufacturing companies. Most of the victims resided in South Korea, India, China, Brazil, Russia, Belarus, Turkey, to name a few.[36] According to the *DPRK's Tactics*, published by the US Army in 2020, the mission of the Lazarus Group is "to create social chaos by weaponizing enemy network vulnerabilities and delivering a payload if directed to do so by the regime."[37] The Bluenoroff Group carries out financial cybercrime by concentrating on long-term assessments and exploiting enemy network vulnerabilities.[38]

[33] Michael Raska, "North Korea's evolving cyber warfare strategy," *East Asia Forum*, September 24, 2020, https://www.eastasiaforum.org/2020/09/24/north-koreas-evolving-cyber-warfare-strategy/.

[34] Jason Bartlett, "Exposing the Financial Footprints of North Korea's Hackers," *Center for a New American Security*, November 18, 2020, https://www.cnas.org/publications/reports/exposing-the-financial-footprints-of-north-koreas-hackers.

[35] U.S. Department of the Treasury, "Treasury Sanctions North Korean State-Sponsored Malicious Cyber Groups," *Press Release*, September 13, 2019, https://home.treasury.gov/news/press-releases/sm774.

[36] Kate Kochetkova, "What is known about the Lazarus Group: Sony hack, military espionage, attacks on Korean banks and other crimes," *Kaspersky Daily*, February 24, 2014, https://www.kaspersky.com/blog/operation-blockbuster/11407/.

[37] U.S. Department of Army, *North Korean Tactics*, APT7-100.2, July 2020, https://fas.org/irp/doddir/army/atp7-100-2.pdf.

[38] Barlett, "Exposing."

Primary targets of APT 37 are located in South Korea: "various industry verticals, including chemicals, electronics, manufacturing, aerospace, automotive, and healthcare."[39] Since at least 2014, APT38 has carried out operations in more than 16 organizations in at least 13 countries, indicating that the group is a large, prolific operation with extensive resources. Especially, in just the publicly reported heists alone, "APT38 has attempted to steal over $1.1 billion dollars from financial institutions."[40] Andarial gathers information through reconnaissance of enemy computer systems and provides initial assessments of enemy networks' vulnerabilities. Initially, it conducted cyber espionage targeting South Korea's military agencies, defence industries, political organizations, security companies, ICT companies, and energy research institutes. Lately, however, its primary targets were shifted to financial targets, such as ATMs, banks, travel agencies, cryptocurrencies, and online gambling users.[41]

3 Cyber Cooperation Between South Korea and the EU

3.1 South Korea's Cyber Strategy

There are two principal documents that guide South Korea's contemporary cyber strategy. The first is the *National Security Strategy (NSS)*, published in December 2018.[42] The *NSS 2018* stipulates that it has three objectives: finding a peaceful solution to the North Korean nuclear issue and establishing permanent peace, contributing to the peace and prosperity of Northeast Asia and the world, and creating a secure society that protects the lives and safety of its people.[43] As one of the action plans to accomplish the second objective, the *NSS 2018* emphasizes that South Korea should proactively

[39] 38 FireEye, "APT37 (Reaper): The Overlooked North Korean Actor," *Threat Research*, February 20, 2018, https://www.fireeye.com/blog/threat-research/2018/02/apt37-overlooked-north-korean-actor.html.

[40] FireEye, "APT38: Details on New North Korean Regime-Backed Threat Group," Threat Research, October 3, 2018, https://www.fireeye.com/blog/threat-research/2018/10/apt38-details-on-new-north-korean-regime-backed-threat-group.html.

[41] Barlett, "Exposing."; Raska, "North," p. 7.

[42] Blue House, *National Security Strategy*, December 2018, (henceforth *NSS 2019*), https://www1.president.go.kr/dn/5c1b2e9a52b8e.

[43] Yeon-cheol Seong, "Blue House publishes national defense strategy of Moon administration," *HanKyoreh*, December 21, 2018.

fulfil its role as a responsible middle power by expanding its contribution to solving global security agendas, including cybersecurity, climate change, and international terrorism.[44] At the same time, it urges South Korea to actively participate in international cooperation to realize universal values of humanity, such as peacekeeping operations, arms control regimes, as well as a common response to cyber threats, to name a few.[45]

In line with the NSS 2018, the *National Cybersecurity Strategy (NCS)* released in April 2019 sets "(creating) a free and safe cyberspace to support national security, promote economic prosperity, and contribute to international peace" as its vision.[46] The *NCS 2019* outlines how South Korea will ensure the society's continuance of reaping benefits and lowering risks deriving from information and communications technologies (ICTs) with six pillars:

1. Secured national critical infrastructure.
2. Enhanced cyber attack defence capabilities.
3. Trust/cooperation-based governance.
4. Cybersecurity industry growth.
5. Well-fostered cybersecurity culture.
6. Strengthened international cooperation.

The three basic principles of the *NSC 2019* are as follows:

1. Balance individual rights with cybersecurity (strike a balance between protecting cyberspace and safeguarding the fundamental rights of the people, e.g., privacy).
2. Conduct security activities based on the rule of law (carry out the government's cybersecurity policies and activities in a transparent manner and in compliance with domestic and international laws).
3. Build a system of participation and cooperation (encourage individuals, businesses, and the government to participate in cybersecurity activities, and pursue close cooperation with the international community).[47]

[44] *NSS*, p. 26.

[45] Ibid., p. 95.

[46] Blue House, *National Cybersecurity Strategy* 2019, (henceforth *NCS 2020*), https://www.itu.int/en/ITU-D/Cybersecurity/Documents/National_Strategies_Repository/National%20Cybersecurity%20Strategy_South%20Korea.pdf.

[47] UNIDIR, "Republic of Korea: Cybersecurity Policy," *Cyber Policy Portal*, September 2020, https://unidir.org/cpp/en/states/republicofkorea.

On the other hand, under the subtitle "Bolstering Response Capabilities against Cyber Threats," the *Defense White Paper 2020*, a biennial publication by the Ministry of Defense,[48] specified several measures to be taken for countering cyber threats:

1. Establishing Cyber-policy and Cyber-strategy and Improving Cyberwarfare Related Preparedness for Mission Accomplishment.
2. Improving Organization and Personnel Management, as well as Education and Training of the Cyber-related Workforce.
3. Enhancing Cyber Capabilities and a Cyber-response System. The MND (Ministry of Defense) plans to further enhance and upgrade its cyber response capabilities to be able to take prompt and systematic actions in response to cyber crises by carrying out cyber wargames on a regular basis which hypothesizes various scenarios in cyberspace, as well as expanding and deepening mutual coordination among and between private and public sectors.
4. Strengthening International Cooperation in Cybersecurity. The MND has been proactive in strengthening international cooperation in the cybersecurity area, in establishing various bilateral and multilateral consultation channels.

Despite the countermeasures against the heightened cyber threats, several loopholes and lacunas in South Korea's overall cyber strategy have been pointed out. First, despite the fact that the National Cyber Security Strategy was touted as the first-ever comprehensive cyber-related government document, its release was so long delayed that this was interpreted by some as "a relatively low prioritization of cybersecurity issues compared to issues, such as inter-Korean relations."[49] This becomes clear when

[48] Ministry of Defense (MND), *Defense White Paper*, 2018, (henceforth, *DWP 2020*), https://www.mnd.go.kr/user/mnd/upload/pblictn/PBLICTNEBOOK_202102021153057640.pdf. (As of early February 2021, English version is not available yet.)

[49] Hannes Ebert and Laura Groenendaal, "Digital Dialogue: Cyber Resilience and Diplomacy in the Republic of Korea: Prospects for EU Cooperation," *EU Cyber Direct*, August 2020, https://eucyberdirect.eu/wp-content/uploads/2020/08/digital-dialogue-rok.pdf.

comparing the US and Japan who had published corresponding documents sixteen and five years earlier respectively.[50]

Second, the highest-standing legal law in the cyber area in South Korea is the National Cybersecurity Management Regulation (NCMR: 국가사이버안전관리규정).[51] The NCMR, promulgated in the form of a Presidential Directive in 2003, is the *only* law that regulates the cybersecurity issue. The problem is that the NCMR, as well as other cyber-related rules and regulations in the national defence sector, does not include cyberspace in the area of responsibility.

Third, the creation of Cyber Operations Command (사이버작전사령부) in August 2018, renamed from Cyber Command (사이버사령부), was a product of domestic political infighting rather than an outcome of forward-looking strategic vision. It reports directly to the Chairman of Joint Chief of Staffs, and therefore only indirectly to the Minister of National Defense.

3.2 South Korea-EU's Strategic Partnership in the Cyber Sector[52]

In a sense, South Korea and the EU are "underappreciated" international security partners. On the one hand, South Korea-EU security relations are anchored in the wider geopolitical context: collective rise of Asia, intensifying power struggle between the US and China, impacting on Hong Kong, Taiwan, South and East China Sea, return of revisionist Russia, the war in the Ukraine, lingering problems in Eastern Europe, including massive refugees and migration from Africa.

On the other hand, South Korea-EU relations are affected by domestic political factors. The EU is faced with the coronavirus pandemic, terrorism fears, renewed Russian aggression at its Eastern borders, a surge in refugees from the Middle East and North Africa, a de facto breakdown of the Schengen system, as well as growing cyber threats.

[50] The White House, *The National Strategy to Secure Cyberspace*, February 2003; Information Security Policy Council of Japan, *Cybersecurity Strategy*, June 10, 2013.

[51] Cybersecurity-related laws have been proposed several times since 2006, but they were often repealed after sluggish discussions. Since the Cyber Security Basic Act has been also pending for over a year, the entire governance of cyber security is still dependent on the presidential decree.

[52] For EU-South Korea cooperation on other areas such as data protection and emerging technologies, see George **Christou** and Ji Soo **Lee** in this volume.

South Korea, on its part, has its plate full of inter-Korean relations, North Korea's nuclear problem, renewed acrimony with Japan, South-South conflict (controversies and debates between conservative and progressive South Korean groups over the policy towards North Korea in general, "남남갈등"), not to mention North Korea's hostile activities in cyberspace. Therefore, cybersecurity poses a common threat which should lead to an intensive exchange of experience leading in turn to common solutions, not least part of a common global strategy. In fact, "a European pivot to Asia" has been underway from the early 2010s not just in economic terms, but also in the political and security spheres.[53]

In the midst of high-stakes and destabilizing geostrategic Sino-US competition across the Indo-Pacific region, South Korea is pursuing a hedging policy, looking beyond Washington and Beijing for other political, economic, strategic, and military partners in order to avoid the danger of being caught in a Sino-US conflict.

Therefore, the EU can be an attractive choice in this regard. Of special interest is growing South Korea-EU cooperation on common issues of cyber threats, as institutionalized by the EU-South Korea Cyber Dialogue (an annual bilateral meeting to address cyberspace), Internet governance, cybersecurity, cyber capacity building in third countries, and cybercrime. Considering that North Korea's cyberwarriors have successfully attacked European institutions including the UK in the 2017 WannaCry ransomware hacking, both sides will benefit from mutual cooperation on cybersecurity.[54]

The security backbone of the bilateral partnership is the *Framework Participation Agreement* signed in 2014.[55] In June 2020, a South Korea-EU joint press statement underlied their commitment "to strengthen cooperation in responding to the Fourth Industrial Revolution, harnessing human-centric digitalization and promoting safe and ethical

[53] S. Boisseau du Rocher and B. Hellendorff, "EU in Asia: Between a Pivot and a Look East," *Diplomat*, July 3, 2014.

[54] Mason Richey, "EU-South Korea Security Relations: The Current State of Play," *Security Policy Brief*, No. 87, May 2017, https://www.egmontinstitute.be/content/uploads/2017/05/SPB87.pdf?type=pdf.

[55] "Agreement between the European Union and the Republic of Korea establishing a framework for the participation of the Republic of Korea in European Union crisis management operations." Date of signature: 23 May 2014. Date of entry into force: December 1, 2016, https://ec.europa.eu/world/agreements/downloadFile.do?fullText=yes&treatyTransId=16823.

implementation of Artificial Intelligence technologies" after a leaders' meeting via video conference in lieu of the planned Summit. They also confirmed the launch of the High-Level Dialogue on digital economy in the near future, discussed the geopolitical situation in the light of the covid-19 pandemic, and affirmed their resolve to uphold the rules-based international order and practical cooperation "in areas such as cybersecurity and disinformation and other hybrid threats."[56]

In fact, since the inception of a full-fledged strategic partnership in 2010, South Korea and the EU have reached a consensus over a set of shared interests related to cyber resilience and stability. First, they are committed to upholding an international rules-based order in cyberspace that complies with existing international public and human rights law. They shared the same view that "multilateral cooperation and a rules based system are needed to guarantee an open, stable and secure cyberspace."[57] Second, setting standards and sharing best practices are essential as both sides have heavily invested in emerging areas of the digital economy such as machine learning, AI, and 5G. Third, the ever-increasing cyber threats create a common interest in sharing best practices on cyber defence and cyber resilience.[58]

4 Conclusion: Policy Recommendations

As demonstrated, South Korea's advanced ICT sector and Internet infrastructure are an obvious and lucrative target for North Korea's cyber attacks. Nevertheless, Seoul's cybersecurity posture and preparedness are unsatisfactory. This is due to the fact that South Korea's cyber policies and strategies are hamstrung by three factors. First, the Korean intelligence apparatus suffers from a deep-rooted and strong negative legacy. For instance, the Korea Central Intelligence Agency (중앙정보부) and the Defense Security Command (국군기무사령부) have been renamed several

[56] European Commission, "Joint Press Release: Republic of Korea – EU Leaders' Video Conference Meeting," *Press Release*, June 30, 2020, https://ec.europa.eu/commission/presscorner/detail/en/IP_20_1226.

[57] High Representative. "Declaration by the High Representative on behalf of the EU on respect for the rules-based order in cyberspace".

Council of EU, April 12, 2019, https://www.consilium.europa.eu/en/press/press-releases/2019/04/12/declaration-by-the-high-representative-on-behalf-of-the-eu-on-respect-for-the-rules-based-order-in-cyberspace/.

[58] Ebert and Groenendaal, "Digital," p. 30.

times due to public resentment and suspicion over their unlawful interference into domestic political affairs. As a consequence, they were often downsized and underfunded. The newly created Cyber Operations Command faces a similar fate. Second, cybersecurity issues were not top on the priority list of the Moon administration. Third, lawmakers do not pay appropriate attention to cybersecurity. They have become somehow lethargic as their efforts to enact a comprehensive legislature often meet with major pushbacks from NGOs and civic groups who strongly advocate privacy and personal freedom.

The following recommendations could help remedying the shortcomings and develop cybersecurity further:

First and foremost, a whole-of-the-government consensus on the need to meet the clear and pressing danger of cybersecurity threat must be established. The massive Russian cyber attack against the US in December 2020 was a security threat of epic proportions and labelled "a wake-up call."[59] South Korea had this wake-up call already earlier. During 2009~2015, there was a series of DDoS attacks against the Presidential Office, the Ministry of National Defense, the National Assembly, and other major government agencies. In addition, the Korea Hydro and Nuclear Power Co Ltd, which runs South Korea's twenty-three nuclear reactors was attacked. In terms of politics, a cyber attack managed to steal OPLAN 5027, the operational procedures for the "decapitation" of Kim Jong-un. Despite these significant and severe attacks, the "awakening moment" of the South Korean government has yet to come.

Second, there is no central authority in charge of overall cybersecurity. The decentralized National Cyber Security System (NCSS) leads to many shortcomings and inefficiencies as the various players, the Ministry of National Defense, the National Intelligence Service, and the Ministry of Security and Public Administration establish and implement their own policies, and strategies. Therefore, "post-incident management for recurrence prevention is difficult to accomplish, which thereby renders inefficient any comprehensive, preventative policymaking."[60] There is a need for inter-agency cooperation and coordination, not silo-thinking. This

[59] Bruce Schneier, "The US has suffered a massive cyberbreach. It's hard to overstate how bad it is," *The Guardian*, December 23, 2020.

[60] Anthony H. Cordesman, Charles Ayers and Aaron Lin, "DPRK: Cyber, Electronic Warfare, and SIGINT Capabilities," *Center for Strategic and International Studies (CSIS)*, 2016, https://www.jstor.org/stable/pdf/resrep23358.7.pdf?refreqid=excelsior%3Aa6f12d1060b61b613571a97ce10fc504.

would be in line with major powers such as the US, Japan, Germany, and the UK which adopted a centralized system addressing cybersecurity at a national level.[61] South Korea's government needs to consolidate its cyber-related strategies and policy. Otherwise, it will not be able to effectively fend off and deter cyberthreats posed by nation-states like North Korea and other cybercriminals.[62] A cabinet-level "Cyber Czar" within the NSC could be a solution. Prioritizing cyber would also send a signal to the US that they are "putting their money where their mouth is."[63] As North Korea is the major cyber attacker, MND should create deputy-minister-level positions to consolidate and integrate overall cyber-related activities, policies, and strategies, at least within the MND.

Fourth, a concept of "defend forward" should be given a thoughtful consideration. The basic idea of this concept is to "going toe-to-toe with adversaries seeking to do (us) harm in cyberspace."[64] According to Paul Nakasone, Commander of the US Cyber Command, the basis of the defend forward strategy is the doctrine of persistent engagement. This is based on the experience that "so much of the corrosive effects of cyberattacks against the United States occur below the threshold of traditional armed conflict. Yet much of Cyber Command's combat power had been devoted toward preparations in the event of future contingencies." Thus, "to do more than prepare for a crisis in the future, it must compete with adversaries today." The goal is to "prevent toeholds from turning into beachheads so that a single compromise will not threaten the military's ability to accomplish its mission."[65]

In analysing the consequences of the weaponization of cyberspace and evaluating its potential direct impact on the security of citizens, South Korea's cybersecurity community started paying attention to the potential

[61] Kim Young-do and others, "Major Issues of the National Cyber Security System in South Korea, and its Future Direction," *Korean Journal of Defense Analysis*, Vol. 25, No. 4 (December 2013): 435–455.

[62] Rothrock, "As Cyberthreats." Ray Rothrock, "As Cyberthreats Grow, Cybersecurity Should Be Centralized," *Governing*, December 20, 2020, https://www.governing.com/news/headlines/As-Cyberthreats-Grow-Cybersecurity-Should-Be-Centralized.html.

[63] Maggie Miller, "Biden's cyber priorities zero in on Russian hack," *Hill*, 27 January 2021.

[64] Maggie Miller, "US Cyber Command Leader Vows to 'Defend Forward' in Protecting Nation from Cyberattacks," *Hill*, August 28, 2020.

[65] Paul M. Nakasone and Michael Sulmeyer, "How to Compete in Cyberspace: Cyber Command's New Approach," *Foreign Affairs*, August 25, 2020, https://www.foreignaffairs.com/articles/united-states/2020-08-25/cybersecurity.

usefulness of the "defend forward" strategy. While the Cyber Operations Command[66] still has to lay out a specific cyber doctrine to implement the "defend forward" strategy, many South Korean experts and scholars have already launched a wide range of studies on relevant topics like deterrence, pre-emptive self-defence, countermeasures, or counterattacks in cyberspace. An important first step could be adding cyberspace to the coverage of the South Korea-US military alliance.

Given the global nature of cybersecurity it is imperative that the South Korean government institutionalizes international cooperation in cybersecurity. One way is building on the existing strategic partnership with the EU. Acceding quickly to the Budapest Convention on Cybercrime could boost multilateral cooperation. The *National Cybersecurity Strategy* stressed the "need to pursue systematic and practical international cooperative activities, such as joining international conventions, sharing information and technology, and drawing up international rules on cybersecurity."[67] The Convention is the first international agreement on cybercrime committed via the Internet and other computer networks.[68] Therefore, the Convention would serve as a crucial platform by which South Korea and the EU could join forces for addressing and dealing with cyber attacks and cybercrime especially committed by North Korea in a coordinated and mutually beneficial way.

References

Bartlett, Jason. "Exposing the Financial Footprints of DPRK's Hackers," *Center for a New American Security*, 18 November 18, 2020. https://www.cnas.org/publications/reports/exposing-the-financial-footprints-of-north-koreas-hackers.

Boisseau du Rocher, and Hellendorff, B. "EU in Asia: Between a Pivot and a Look East," *Diplomat*, July 3, 2014.

[66] In 2019, South Korea overhauled the country's cyber command, renaming the unit the Cyber Operations Command and designating it as an entity under the direct control of the chairman of the Joint Chiefs of Staff for the purpose of reinvigorating its military operational focus.

[67] *NCS 2019*, p. 9.

[68] Council of Europe, "Details of Treaty No.185: Convention on Cybercrime," November 23, 2001, https://www.coe.int/en/web/conventions/full-list/-/conventions/treaty/185.

Biden, Joe. "The Power of America's Example: The Biden Plan for Leading the Democratic World to Meet the Challenges of the 21st Century", 2020. https://joebiden.com/americanleadership/.

Blue House, the Republic of Korea. *National Security Strategy*, December 2018. https://www1.president.go.kr/dn/5c1b2e9a52b8e.

Blue House, the Republic of Korea. *National Cybersecurity Strategy*. 2019. https://www.itu.int/en/ITU-D/Cybersecurity/Documents/National_Strategies_Repository/National%20Cybersecurity%20Strategy_South%20Korea.pdf.

Chong-woo Kim and Carolina Polito. "The Evolution of North Korean Cyber Threats," *Asan Institute for Policy Studies*, February 20, 2019. http://en.asaninst.org/contents/the-evolution-of-north-korean-cyber-threats/.

CISA (*Cybersecurity and Infrastructure Security Agency*). "Alert (AA20-106A): Guidance on the DPRKn Cyber Threat," April 15, 2020. https://us-cert.cisa.gov/ncas/alerts/aa20-106a.

Cordesman, Anthony H. Charles Ayers and Aaron Lin. "DPRK: Cyber, Electronic Warfare, and SIGINT Capabilities," Center for Strategic and International Studies (CSIS), 2016. https://www.jstor.org/tc/accept?origin=%2Fstable%2Fpdf%2Fresrep23358.7.pdf%3Frefreqid%3Dexcelsior%253Aa6f12d1060b61b613571a97ce10fc504&is_image=False.

Council of Europe. (2001). "Details of Treaty No.185: Convention on Cybercrime," 23 November. https://www.coe.int/en/web/conventions/full-list/-/conventions/treaty/185.

Ebert, Hannes and Laura Groenendaal. "Digital Dialogue: Cyber Resilience and Diplomacy in the Republic of Korea: Prospects for EU Cooperation," *EU Cyber Direct*, August 2020. https://eucyberdirect.eu/wp-content/uploads/2020/08/digital-dialogue-rok.pdf.

European Commission. "Joint Press Release: Republic of Korea – EU Leaders' Video Conference Meeting," *Press Release*, June 30, 2020. https://ec.europa.eu/commission/presscorner/detail/en/IP_20_1226.

FireEye. "APT37 (Reaper): The Overlooked DPRKn Actor," *Threat Research*, February 20, 2018a. https://www.fireeye.com/blog/threat-research/2018/02/apt37-overlooked-north-korean-actor.html.

FireEye. "APT38: Details on New DPRKn Regime-Backed Threat Group," *Threat Research*, October 3, 2018b. https://www.fireeye.com/blog/threat-research/2018/10/apt38-details-on-new-north-korean-regime-backed-threat-group.html.

Information Security Policy Council of Japan. *Cybersecurity Strategy*, June 10, 2013.

Ioanes, Ellen. "Kim Jong Un has quietly built a 7,000-man cyber army that gives DPRK an edge nuclear weapons don't," *Business Insider*, July 17, 2020.

Jun, Jenny et al. "The Organization of Cyber Operations in DPRK," *CSIS (Korea Chair Platform)*, 18 December 18, 2014. https://csis-website-prod.s3.amazonaws.com/s3fs-public/legacy_files/files/publication/141218_Cyber_Operations_North_Korea.pdf.

Jun, Jenny et al. *DPRK's Cyber Operations: Strategy and Response*, New York: CSIS, 2015. http://csis-website-prod.s3.amazonaws.com/s3fs-public/legacy_files/files/publication/151216_Cha_NorthKoreasCyberOperations_Web.pdf.

Kim, Young-do et al. "Major Issues of the National Cyber Security System in South Korea, and its Future Direction," *Korean Journal of Defense Analysis*, Vol. 25, No. 4 (December 2013), pp. 435–455.

Ko, Kyungmin et al. "The Internet dilemma and control policy: political and economic implications of the Internet in DPRK," *Korean Journal of Defense Analysis*, Vol. 21, No. 3 (September 2009), pp. 279–295.

Kochetkova, Kate. "What is known about the Lazarus Group: Sony hack, military espionage, attacks on Korean banks and other crimes," *Kaspersky Daily*, February 24, 2014. https://www.kaspersky.com/blog/operation-blockbuster/11407/.

Kong, Ji-young et al. "The All-Purpose Sword: DPRK's Cyber Operations and Strategies," *11th International Conference on Cyber Conflict (CyCon)*, Tallinn, Estonia,2019https://ccdcoe.org/uploads/2019/06/Art_08_The-All-Purpose-Sword.pdf.

Lederer, Edith. "UN probing 35 DPRKn cyberattacks in 17 countries," *AP News*, August 13, 2019.

Lee, Young-jong. "Cyber warfare is a merciless treasure-sword of the Korean People's Army," *Sisa Journal*, December 30, 2014. http://www.sisajournal.com/news/articleView.html?idxno=140623.

Microsoft. *Microsoft Digital Defense Report*, September 2020. https://www.microsoft.com/en-us/download/confirmation.aspx?id=101738.

Miller, Maggie. "US Cyber Command Leader Vows to 'Defend Forward' in Protecting Nation from Cyberattacks," *Hill*, August 28, 2020a.

Miller, Maggie. "Microsoft Reports Spike in Foreign Targeting of COVID-19 Researchers, Think Tanks," *Hill*, September 29, 2020b.

Miller, Maggie. "Biden's cyber priorities zero in on Russian hack," *Hill*, January 27, 2021.

Miller, Steve. "Where Did DPRK's Cyber Army Come From?," *VOA News*, November 20, 2018.

Ministry of Defense (MND), The Republic of Korea. *Defense White Paper*, 2016. https://www.mnd.go.kr/user/mnd/upload/pblictn/PBLICTNEBOOK_201705180311469090.pdf.

Ministry of Defense (MND). *Defense White Paper*. 2018. https://www.mnd.go.kr/user/mnd/upload/pblictn/PBLICTNEBOOK_202102021153057640.pdf.

Nakasone, Paul M. and Michael Sulmeyer. "How to Compete in Cyberspace: Cyber Command's New Approach," *Foreign Affairs*, August 25, 2020. https://www.foreignaffairs.com/articles/united-states/2020-08-25/cybersecurity.

Palmer, Annie. "Coupang, a SoftBank-backed start-up, is crushing Amazon to become South Korea's biggest online retailer," *CNBC*, June 16, 2020.

Park, Ju-min and James Pearson. "Exclusive: DPRK's Unit 180, the cyber warfare cell that worries the West," *Reuters*, May 21, 2017.

Raska, Michael. "DPRK's evolving cyber warfare strategy," *East Asia Forum*, September 24, 2020. https://www.eastasiaforum.org/2020/09/24/north-koreas-evolving-cyber-warfare-strategy/.

Recorded Future. "DPRK Cyber Activity," *Recorded Future Insikt Group Report*, June 14, 2020. https://go.recordedfuture.com/hubfs/reports/north-korea-activity.pdf.

High Representative. "Declaration by the High Representative on behalf of the EU on respect for the rules-based order in cyberspace", Council of EU, April 12, 2019. https://www.consilium.europa.eu/en/press/press-releases/2019/04/12/declaration-by-the-high-representative-on-behalf-of-the-eu-on-respect-for-the-rules-based-order-in-cyberspace/.

Reporters Without Borders. "Internet Enemies 2011 – DPRK," March 11, 2011. https://www.refworld.org/docid/4d82268ec.html.

Richey, Mason. "EU-South Korea Security Relations: The Current State of Play," *Security Policy Brief*, No. 87, May 2017a. https://www.egmontinstitute.be/content/uploads/2017/05/SPB87.pdf?type=pdf.

Richey, Mason. "EU-South Korea Security Relations: The Current State of Play," *Security Policy Brief*, No. 87, May 2017b. https://www.egmontinstitute.be/content/uploads/2017/05/SPB87.pdf?type=pdf.

Rothrock, Ray. "As Cyberthreats Grow, Cybersecurity Should Be Centralized," *Governing*, December 20, 2020. https://www.governing.com/news/headlines/As-Cyberthreats-Grow-Cybersecurity-Should-Be-Centralized.html.

Sanger, David E. and Nicole Perlroth. "U.S. Accuses DPRK of Cyberattacks, a Sign That Deterrence Is Failing," *New York Times*, April 15, 2020.

Sanger, David E. et al. "The World Once Laughed at North Korean Cyberpower. No More," *New York Times*, October 15, 2017. https://www.nytimes.com/2017/10/15/world/asia/north-korea-hacking-cyber-sony.html.

Sanger, David E. et al. "Billions Spent on U.S. Defenses Failed to Detect Giant Russian Hack," *New York Times*, December 16, 2020. https://www.nytimes.com/2020/12/16/us/politics/russia-hack-putin-trump-biden.html.

Schneier, Bruce. "The US has suffered a massive cyberbreach. It's hard to overstate how bad it is," *The Guardian*, December 23, 2020.

Seong, Yeon-cheo. "Blue House publishes national defense strategy of Moon administration," *HanKyoreh*, December 21, 2018.

Tharoor, Ishaan. "An Emerging New Alliance of Democracies," *Washington Post*, December 18, 2020.

U.S. Department of Army. *North Korean Tactics*, APT 7-100.2, July 2020. https://fas.org/irp/doddir/army/atp7-100-2.pdf.

U.S. Department of the Treasury. "Treasury Sanctions North Korean State-Sponsored Malicious Cyber Groups," *Press Release*, September 13, 2019. https://home.treasury.gov/news/press-releases/sm774.

U.S. Government Accountability Office (GAO). "Information Technology: Federal Agencies Need to Take Urgent Action to Manage Supply Chain Risks," *GAO-21-171*, December 2020. https://www.gao.gov/assets/720/711266.pdf.

UNIDIR. "Republic of Korea: Cybersecurity Policy," *Cyber Policy Portal*, September 2020. https://unidir.org/cpp/en/states/republicofkorea.

Venard, Bertrand. "The Cold War 2.0 between China and the US Is Already a Virtual Reality," *Conversation*, October 17, 2019.

Waldeck, Yasmin. "Smartphone penetration rate as share of the population in South Korea from 2015 to 2025," *Statista*, July 17, 2020. https://www.statista.com/statistics/321408/smartphone-user-penetration-in-south-korea/.

White House. *The National Strategy to Secure Cyberspace*, February 2003. https://us-cert.cisa.gov/sites/default/files/publications/cyberspace_strategy.pdf.

Williams, Martyn. (2016). "How the Internet Works in North Korea," *Slate*, November 28, 2016. https://slate.com/technology/2016/11/how-the-internet-works-in-north-korea.html.

Building Cyber Resilience: The Defensive Shield for the EU

Anna-Maria Osula

1 Introduction

The digital transformation of our societies has expanded the attack surface, rendering the integrity, availability, and confidentiality of data an issue of strategic importance. The number of malicious cyber incidents is on the rise, their nature is becoming more sophisticated, and attribution, together with increasingly advanced adversary capabilities, continues to be

Research for this chapter was supported by the Masaryk University ERDF project "CyberSecurity, CyberCrime and Critical Information Infrastructures Center of Excellence" (No. CZ.02.1.01 / 0.0 / 0.0 / 16_019 / 0000822).

A.-M. Osula (✉)
Tallinn University of Technology, Tallinn, Estonia

Guardtime, Tallinn, Estonia

Masaryk University, Brno, Czechia

© The Author(s), under exclusive license to Springer Nature Switzerland AG 2022

G. Boulet et al. (eds.), *Cybersecurity Policy in the EU and South Korea from Consultation to Action*, New Security Challenges,
https://doi.org/10.1007/978-3-031-08384-6_9

a challenge.[1] These attacks are originating from different actors globally, at times showcasing political motives, and thereby possibly threatening democratic processes such as elections. Cyber incidents can be particularly damaging if targeted against critical infrastructure and essential services.

On top of that, there are global tensions over the control of technologies across the whole supply chain as well as new computing and trusted data processing paradigms such as blockchain, cloud-to-edge, and quantum computing.[2] Another critical issue for Europe's industrial and technological base is being largely dependent on having reliable access to raw materials from the international markets and to the necessary foreign components and technologies. Compared to China and the US, the EU also falls behind in the production of key digital technologies such as AI and its public investment in research and development programmes.[3] Being dependent on other actors decreases the level of scrutiny over the security of these key technologies.

Despite all these concerns, the EU remains positive regarding its central role on the global arena and confirms its ambition to be in the lead for digital economy, invest more in technology, and remain the frontrunner in maintaining a high level of protection for the whole society. In order to reach these goals, EU has identified a number of measures in its new Cybersecurity Strategy. These can be divided into three: (1) resilience, technological sovereignty, and leadership; (2) building operational capacity to prevent, deter, and respond; and (3) advancing a global and open cyberspace.[4] The strategy reveals that a big part of the EU's projected success as a digital leader is the development and effective use of new technologies. This would give the EU an edge in the global market, drive growth and innovation, and protect its citizens, private and public sector

[1] ENISA, 'ENISA Threat Landscape – The Year in Review', 2020, 8, https://www.enisa.europa.eu/publications/year-in-review.

[2] European Commission, 'Joint Communication to the European Parliament and the Council: The EU's Cybersecurity Strategy for the Digital Decade (JOIN (2020)18 Final)', 2020, 1.

[3] European Political Strategy Centre, 'Rethinking Strategic Autonomy in the Digital Age', EPSC Strategic Notes, 2019, 8–10, https://op.europa.eu/en/publication-detail/-/publication/889dd7b7-0cde-11ea-8c1f-01aa75ed71a1/language-en/format-PDF/source-118064052.

[4] European Commission (2020b).

alike. Equally, the European Commission has stated that for the next five years "technological sovereignty" will be at the centre of its strategy.[5]

However, while the EU has worked hard on establishing solid policies and regulation for ensuring cybersecurity in general, its overall stance in defence-related matters has for a long time been not so clear-cut. In particular, while the EU has identified that its security faces threats such as state and non-state actors targeting the EU with hybrid tools, including disruptive technologies, disinformation, and other non-military sources of influence, it took its time for developing a policy to guide its actions in a geopolitically contested environment.[6] In March 2022, underlining the need for an accelerated response for geopolitical challenges, the EU adopted the Strategic Compass which demonstrates the member states' common vision about the EU as a security provider. Importantly, the strategy calls for, inter alia, further developing the EU Cyber Defence Policy to allow for better preparedness for and response to cyberattacks, and includes a number of actions tailored for boosting cyber defence research and education, cooperation, capability development and generally enhancing EU's cyber posture.

This chapter will first explain in a nutshell the EU's main regulatory and policy aims in achieving increased resilience in cyberspace. The chapter then focuses on the role of disruptive technologies in building a defensive shield for the EU and recent developments within the domain of EU cyber defence. Finally, the chapter will illustrate how cooperation with trusted partners such as South Korea will be mutually beneficial for enhancing cyber resilience.

2 Bolstering Cyber Resilience

Cyber resilience and the security of technology go hand in hand with the development of digital products and solutions, use of data, and the ensuing digital transformation of our society. While cybersecurity describes an

[5] Ursula von der Leyen, 'Political Guidelines for the Next European Commission 2019–2024', 2019, 13, https://ec.europa.eu/commission/sites/beta-political/files/political-guidelines-next-commission_en.pdf.

[6] European Commission, 2021, Strategic Compass, https://eeas.europa.eu/sites/default/files/towards-a-strategic-compass-2021-february.pdf; see also: EUISS, Finding direction with a Strategic Compass? Reflections on the future of EU security and defence, 2020, 2, https://www.iss.europa.eu/sites/default/files/EUISSFiles/Event%20Report%20-%20Strategic%20Compass.pdf.

actor's ability to protect against and prevent cyber threats, resilience refers to the ability to mitigate the damage of an incident and continue operation. In today's setting, both of these elements—cybersecurity and cyber resilience—are relevant for the functioning of our societies, and allow for people to trust, use, and benefit from innovation, connectivity, and automation.

It is important to underline that cyber resilience is not a separate area of regulation or policy, and cannot be viewed in a silo. Instead, its development is, and must be, interlinked with the policy and regulation of other domains such as digitisation, digital identity, privacy, data protection, cybercrime, and the safety and security of our societies. It is also worth to point out that building cyber resilience is not anymore an issue merely on domestic agendas, but requires cooperation between the EU member states, among the EU institutions and the multi-stakeholder community and international partners.

The EU has been working on enhancing cyber resilience for a while. Already in 2001, the European Commission adopted the Communication on Network and Information Security which highlighted the importance of the ability of networks and information systems to resist accidental events or malicious actions.[7] Fast-forward to 2013, and the EU listed reaching cyber resilience as the primary goal of the first EU Cybersecurity Strategy.[8] In 2017, the EU put forward an approach to cybersecurity with resilience-building, rapid response, and effective deterrence at its core,[9] and underlined again the centrality of cyber resilience in the 2020 Cybersecurity Strategy. Some of the most relevant related legal agreements are the General Data Protection Regulation (GDPR), the EU Network and Information Security Directive (NIS Directive), and the Cybersecurity Act. There are also several institutional arrangements in place, such as the EU-level computer emergency response team CERT-EU,

[7] European Commission, 'Communication from the Commission to the European Parliament, the Council, the European Economic and Social Committee and the Committee of the Regions – Network and Information Security: Proposal for A European Policy Approach (COM(2001)298)', 2001.

[8] European Commission, 'Communication from the Commission to the European Parliament, the Council, the European Economic and Social Committee and the Committee of the Regions – Cybersecurity Strategy of the European Union: An Open, Safe and Secure Cyberspace (JOIN(2013) 1 Final)', 2013.

[9] Joint Communication on Resilience, Deterrence and Defence: Building strong cybersecurity for the EU, JOIN (2017) 450.

the European Network and Information Security Agency, the European Defence Agency, and the European Cybercrime Centre within Europol.

2.1 Resilience and Technological Sovereignty

The EU sees cyber resilience as a key capability in achieving and maintaining technological sovereignty. This includes ensuring that all Internet-connected devices have been designed keeping security in mind, operationally recover from cyber incidents, and are quickly patched when vulnerabilities are discovered.[10] One of the main regulatory tools to achieve this goal is the NIS Directive,[11] adopted in 2016. By providing legal measures to boost and harmonise the overall level of cybersecurity in the EU, the Directive focuses on increasing member states' preparedness for cyber incidents, cooperation among member states, and introducing a CSIRT Network with the aim to promote swift and effective operational cooperation on cybersecurity incidents and sharing information about vulnerabilities and risks.

Applying the NIS Directive domestically has outlined several challenges. Firstly, there is a low level of cyber resilience of private entities operating in the EU. Secondly, the level of cyber resilience across member states and sectors is inconsistent. Thirdly, there appears to be a low level of joint situational awareness and lack of joint crisis response. Therefore, the upcoming NIS Directive review aims to, *inter alia*, harmonise security and incident reporting requirements, align national authorities and enforcement rules, and level the capabilities of competent agencies across the EU.[12]

Another limitation of the NIS Directive is establishing security requirements for "essential services". This means that the Directive has left uncovered many other relevant digital and physical infrastructure and services. In responding to this shortcoming, the European Commission (EC) has proposed a Directive on the Resilience of Critical Entities, which underlines the importance of protecting critical services from all types of

[10] European Commission (2020b, 5).

[11] European Union, 'The Directive (EU) 2016/1148 of the European Parliament and of the Council of 6 July 2016 Concerning Measures for a High Common Level of Security of Network and Information Systems across the Union', 2016.

[12] European Commission, 'Proposal for a Directive of the European Parliament and of the Council on Measures for a High Common Level of Cybersecurity across the Union, Repealing Directive (EU) 2016/1148 (COM(2020) 823 Final)', 2020, 1.

attacks, not only cyber.[13] Such regulation is critical for enhancing EU's cyber resilience because of the possible implications malicious incidents related to critical entities may have with domestic and EU security. Defending critical entities has become a priority as their operators are confronted with challenges related to integrating new technologies such as 5G and unmanned vehicles into their operations, while at the same time having difficulties in fully addressing the vulnerabilities that such technologies could potentially bring along. Equally, the use of new technologies may make operators increasingly reliant on one another, as well as dependent on specific technological solutions.[14]

These efforts go hand in hand with the promotion of the concept of "security by design" which envisages widespread awareness of cybersecurity threats and building security into processes, products, and services right from the start. To facilitate this process, the EU is working on cybersecurity certification schemes and develops standards, with the purpose to ensure an adequate level of cybersecurity as well as reducing the fragmentation of the internal market.[15]

In addition, the EC has proposed a number of new initiatives aimed at boosting EU's resilience. For example, in order to create collective knowledge of cybersecurity incidents, a network of Security Operations Centres will be established across the EU, serving as a real "Cyber Shield" for the EU. These centres will be able to improve the speed of incident detection, analysis, and response with the help of AI, machine learning, as well as supercomputing infrastructure developed in the EU by the European High-Performance Computing Joint Undertaking.[16] The EC has also expressed its intent to contribute to securing Internet connectivity by supporting the development of a public European DNS resolver service,

[13] European Commission, 'The Commission Proposes a New Directive to Enhance the Resilience of Critical Entities Providing Essential Services in the EU', Migration and Home Affairs – European Commission, 16 December 2020, https://ec.europa.eu/home-affairs/news/commission-proposes-new-directive-enhance-resilience-critical-entities-providing-essential_en.

[14] European Commission, Proposal for a Directive of the European Parliament and of the Council on Measures for a High Common Level of Cybersecurity across the Union.

[15] European Union, 'Regulation (EU) 2019/881 of the European Parliament and of the Council of 17 April 2019 on ENISA (the European Union Agency for Cybersecurity) and on Information and Communications Technology Cybersecurity Certification and Repealing Regulation (EU) No 526/2013', 2019.

[16] Read more about the European High-Performance Computing Joint Undertaking: https://ec.europa.eu/digital-single-market/en/eurohpc-joint-undertaking for explanation.

which would address the potential vulnerability entailed in the current state of affairs, where the DNS resolution is consolidated in the hands of few companies. Similarly, the EU supports and accelerates the uptake of key Internet standards including IPv6.[17]

2.2 Operational Capacity to Prevent, Deter, and Respond

One of the challenges the EU is facing in ensuring EU's cybersecurity is the cooperation between different communities working on various aspects of cybersecurity. Within the EU, there are four domains: (1) NIS authorities such as CSIRTs and disaster response, (2) law enforcement and judicial authorities, (3) cyber diplomacy, and (4) cyber defence—each with their own set of policy and regulatory tools. The key will be to ensure smooth cooperation between these communities in reaching a comprehensive and inclusive view of EU's cybersecurity threats, goals, and solutions. The communities need to work more closely together and with the member states, share information on cybersecurity awareness, as well as be able to effectively prevent and respond to cyber incidents.[18] While these four domains previously developed to a large extent in silos—cybersecurity was portrayed as a separate domain—by now the focus has shifted to the ability to deliver a coordinated response.[19]

One of the key improvements proposed is the creation of a Joint Cyber Unit (JCU) which would serve as a virtual and physical platform for cooperation for these different cybersecurity communities and play a big role on boosting the EU cyber resilience. The JCU's main goal will be operational and technical coordination against major cross border cyber incidents and threats, as well as establishing structured partnerships with a trusted industry base and facilitating a coordinated approach to cooperation with external partners. The JCU can ensure effective response to cyber threats using collective resources and expertise. Importantly, the JCU would not be an additional, standalone body (as there are many cybersecurity-related bodies already in the EU), nor would it affect the competences and powers of national cybersecurity authorities or EU participants. Rather, the JCU would act as a supporting platform where the

[17] European Commission, 2020b, 7–12.

[18] See the chapter of Gibum **Kim** outlining the difficulties faced in Korea for a common approach.

[19] European Commission (2020b, 13).

participants can make use of each other's expertise, especially when different EU cyber communities are required to work closely together.[20]

Foremost, for the JCU to fulfil its potential, the Union should focus on increasing trust between the member states, as one of the key challenges is currently the lack of systematic information sharing with each other, with negative consequences for the effectiveness of the cybersecurity measures and for the level of joint situational awareness at EU level.[21] Another measure for enhancing information sharing is the establishment of a member states' EU cyber intelligence working group, as well as setting up the Military CERT-Network under the auspices of the European Defence Agency. Regrettably, Brexit adds another layer on the information sharing challenge, as there may be further tensions in the UK-EU relationship on cybersecurity, particularly in reference to balancing the need for intelligence-gathering with EU privacy and data protection concerns.[22]

2.3 *Cyber Defence and Disruptive Technologies*

While security and defence policy in the EU is predominantly a competence of the member states, there is an evolving consensus that the instability and unpredictability characterising the global security environment requires a common response from the EU and its member states.[23] Preventing, establishing an effective response for, and recovering from hybrid threats remain one of the key challenges for the EU. This points to the increasing role of the EU in providing a multidisciplinary and integrated response to security threats, as well as leading the development and harmonisation of member states' capabilities. In particular, the EU needs to focus on strengthening a common European security and defence culture, improving the Union's operational effectiveness, resilience, capabili-

[20] European Commission (2020b, sec. 2.1).

[21] European Commission, 'Proposal for a Directive of the European Parliament and of the Council on Measures for a High Common Level of Cybersecurity across the Union, Repealing Directive (EU) 2016/1148 (COM(2020) 823 Final)', 5.

[22] James Black et al., 'Defence and Security after Brexit: Understanding the Possible Implications of the UK's Decision to Leave the EU – Compendium Report', 130.

[23] Elena Lazarou and Alina Dobreva, 'Security and Defence', European Parliamentary Research Service, 2019, 8. See also European Commission, 'Communication from the Commission to the European Parliament, the Council, the European Economic and Social Committee and the Committee of the Regions on the EU Security Union Strategy (COM/2020/605 Final)', 2020; European Union, 'The European Union's Global Strategy: Three Years on, Looking Forward', 2019.

ties, and cooperation with partners, as well as defining objectives and concrete goals for its policies.[24] The 2020 EU Security Union Strategy reiterates that in facing contemporary complex security threats, EU citizens cannot be protected only through member states acting on their own.[25] The same insight is reflected in the new Cybersecurity Strategy.

The updated **EU Cyber Defence Policy Framework (CDPF)** identified in 2018 priority areas for cyber defence, and called for building stronger cybersecurity, referring to allowing for more measures in responding and deterring cyber attacks.[26] In 2020, the EC has called for the review of the CDPF and supports initiatives such as the upcoming "Military Vision and Strategy on Cyberspace as a Domain of Operations" by the EU Military Committee. While the EU has identified cyberspace as a domain for operations already in 2018,[27] further discussions are needed on how the new domain and capabilities deriving from therein enable the military missions and operations of the EU Common Security and Defence Policy. Likewise, a joint civil-military doctrine for complex emergencies would facilitate streamlining member states' common goals and regular joint civil-military exercises help to test procedures, measures, and lines of command and communication.[28]

In particular, the EU and member states are expected to enhance the development of state-of-the-art cyber defence capabilities through different EU policies and instruments, focusing on the development and use of key technologies such as AI, encryption, and quantum computing.[29] This refers to the invaluable role of disruptive technologies in developing cyber defence capabilities, forming part of the EU's general policy objective of achieving technological sovereignty. The mastery of such frontier technologies will be an essential skillset for the twenty-first-century battlefield, translating into concrete advantages in future conflicts, cyber responsive

[24] European Commission (2021), Strategic Compass; see also: EUISS, 2.

[25] European Commission, 'Communication from the Commission to the European Parliament, the Council, the European Economic and Social Committee and the Committee of the Regions on the EU Security Union Strategy (COM/2020/605 Final)', 1.

[26] Council of the European Union, 'EU Cyber Defence Policy Framework (2018 Update) (14413/18)', 2018.

[27] Ibid.

[28] European Union Institute for Security Studies, 3–4.

[29] European Commission (2020b, 18).

operations and information superiority.[30] However, research shows that countries' approach to cyber defence differs, with some countries opting for more active defence[31] including the openly discussed development of offensive capabilities, while others preferring a more defensive stance.[32]

To that end, the EU is committed to continuing to invest in blockchain, high-performance computing, quantum computing, developing algorithms and tools to allow data sharing and data usage, and believes that data and AI are ingredients for innovation that can help to find solutions to societal challenges ranging from health to manufacturing and defence.[33] For example, blockchain technology has the potential to mitigate these concerns by providing anticipation of possible malign abuses of technologies[34] such a ensuring data and process integrity which may prove especially useful in the context of large datasets and critical services.[35] Blockchain technologies can thus be successfully employed to enforce the concepts of "security by design" and "autonomy by design"[36] which will lay the foundation for any secure technological solution. Moreover, blockchain technologies may also prove crucial to enhance the security of the disruptive technologies themselves.[37]

[30] European Political Strategy Centre, 'Rethinking Strategic Autonomy in the Digital Age', 5. See also European Union, Strategic Compass, 45.

[31] See the chapter of Kyu-dok **Hong** and Seong-jong **Song**, who discuss "defending forward" in South Korea.

[32] Alessandro Marrone and Ester Sabatino, Cyber Defence in NATO Countries: Comparing Models, Instituto Affari Internazionali, 2021, https://www.iai.it/sites/default/files/iaip2105.pdf.

[33] von der Leyen, 'Political Guidelines for the Next European Commission 2019–2024', 13.

[34] European Political Strategy Centre, 'Rethinking Strategic Autonomy in the Digital Age', 2.

[35] Luukas Ilves and Anna-Maria Osula, 'The Technological Sovereignty Dilemma – and How New Technology Can Offer a Way Out', *European Cybersecurity Journal* 6, no. 1 (2020): 30–31. See also Anna-Maria Osula, 'The Global Rush for Standards in Blockchain', *Directions Blog*, 2019, https://directionsblog.eu/the-global-rush-for-standards-in-blockchain/.

[36] Ilves and Osula, 'The Technological Sovereignty Dilemma – and How New Technology Can Offer a Way Out', 29–32.

[37] For example, the EU foresees that cybersecurity must be integrated into all digital investments, particularly key technologies like artificial intelligence, encryption, and quantum computing, using incentives, obligations, and benchmarks. See, European Commission (2020b, 5).

Equally, the EU sees great promise in developing AI solutions for both civilian and military use.[38] AI is expected to help overcoming the challenges related to the use of big data such as the volume and variety of data. Such cutting-edge technological solutions may be highly beneficial for improving situational awareness and decision-making as well as supporting planning, including modelling and simulation. In the defence context, AI has the potential to, for example, assist operators in their understanding of an operational situation by enabling a more advanced decision-making process, and in the optimisation of the use of their assets.[39] Similar use of the AI technology can be seen in the Security Operations Centres discussed in Sect. 2.1.

Member states are encouraged to take advantage of the existing capability development processes under the European Defence Fund or Permanent Structured Cooperation. In order to support defence-related R&D, innovation, and capability development, the EU adopted the European Defence Industrial Development Programme (EDIDP) in 2019. The work programme dedicated €500 million for the European capability development in 2019–2020.[40] The EDIDP includes several projects engaging the latest technological solutions. For example, the EU identified as one of the priorities to develop information management and superiority and command, control, communication, computers, intelligence, surveillance and reconnaissance, cyber defence, and cybersecurity. This includes specific focus on cyber situational awareness and defence capabilities, defence networks and technologies for secure communication and information sharing, for example, for the use of cyber rapid reaction

[38] See, for civilian use, e.g. Communication from the Commission to the European Parliament, the Council, the European Economic and Social Committee and the Committee of the Regions, Fostering a European approach to Artificial Intelligence, COM(2021) 205 final, 21.04.2021. For an overview on the defence setting, see, e.g. Soare, Simona R, European Defence and AI: Game-Changer or Gradual Change?, 24 March 2021, https://www.rsis.edu.sg/rsis-publication/rsis/european-defence-and-ai-game-changer-or-gradual-change/#.YIPvRqkzY-S.

[39] European Commission, European Defence Industrial Development Programme, 2020 calls for proposals, conditions for the calls and annexe, 2020, pp. 93–94.

[40] European Commission, "Annexes to the Commission Implementing Decision on the Financing of the European Defence Industrial Development Programme and the Adoption of the Work Programme for the Years 2019 and 2020", C(2019) 2205 final, Brussels, March 19, 2019, https://ec.europa.eu/docsroom/documents/34515.

teams;[41] Space Situational Awareness and early warning capabilities; and maritime surveillance capabilities. The EDIDP also includes calls on defence technologies supported by AI as well as innovative and future-oriented defence solutions.[42] One of the projects which addresses these concerns is the development of the European Cyber Situational Awareness Platform which will offer integrated and real-time imaging of cyber threats to which a defence system may be subjected. The solution will enable immediate reaction capabilities, as well as smart support tools for decision-making.[43]

3 Global Resilience Through Cooperation

The EU continues to value international and regional cooperation in keeping cyberspace global, open, stable, and secure. Cooperation in cybersecurity questions should not be confined to only specific cybersecurity negotiations but be part of agreeing on policies about economic affairs, domestic security and foreign security, and defence. As a strong economic and trading bloc the EU is uniquely placed to lead the way in defining and upholding international norms and technical standards, supporting the fight against cybercrime, promoting confidence-building measures as well as capacity building.[44]

The EU has always appreciated strong partnerships with likeminded actors as well as regional and international organisations. It therefore aims to strengthen and expand multi-stakeholder partnerships and continue its cyber dialogues with third countries to promote its values and vision for cyberspace, sharing best practices, and seeking to cooperate more effectively. Through supporting global initiatives such as the Budapest Convention on Cybercrime and more coordinated capacity building, the EU commits to strengthening global capacities in addressing cyber threats in partner countries, thereby aiming to increase global resilience.[45]

In the domain of cyber defence in particular, the EU values the close cooperation with the NATO. As cyber attacks have the potential to be as

[41] PESCO Projects, Cyber Rapid Response Teams and Mutual Assistance in Cyber Security, https://pesco.europa.eu/project/cyber-rapid-response-teams-and-mutual-assistance-in-cyber-security/.

[42] European Commission, European Defence Industrial Development Programme, 6–7.

[43] Jack Richardson, 'EDIDP Wins for Indra', *European Security & Defence* (blog), 6 July 2020, https://euro-sd.com/2020/07/articles/18075/edidp-wins-for-indra/.

[44] European Commission (2020b, 20).

[45] European Commission (2020b, 21–22).

harmful as conventional attacks, cyber defence has been recognised by NATO as a domain of operations in which NATO must defend itself as effectively as it does in the air, on land, and at sea. With stating that cyber defence is a part of NATO's core task of collective defence, NATO member states have sent a clear message regarding the priority of enhancing domestic as well as NATO defence capabilities, integrating cyber defence to operational planning and Alliance operations and missions. EU and NATO benefit from cooperation in complementary defence planning, information sharing, joint exercises, and sharing best practices.

As previously iterated, technological sovereignty does not mean isolation or withdrawal from the international stage. Quite the opposite, building and maintaining robust international partnerships, furthering strategic cooperation and capacity building will strengthen the EU's cyber resilience and be beneficial for all partners. Indeed, new technologies offer opportunities not only to become technologically more sovereign but also in preventing, protecting, and mitigating cyber threats in general, and therefore being one of the foundational layers for effective global resilience.

To this end, this author welcomes the European Council's decision to deepen the cooperation with EU's Asian Strategic Partners: China, India, Japan, and South Korea in several security domains.[46] *Inter alia*, these include cybersecurity, counter terrorism, hybrid threats, and conflict prevention. The chapter suggests that technological sovereignty and the deployment of disruptive technologies for "security by design" and "autonomy by design" should be in the agenda of the negotiations with the Asian partners.

Given the complex global geopolitical setting, there is a need for the EU's security and defence cooperation in and with Asia to become operational.[47] Therefore South Korea and the EU, which celebrated in 2020 the tenth anniversary of their strategic partnership, will engage in a deeper cooperation on cybersecurity, adding to the many other areas of enhanced cooperation. In addition to strong trade relations,[48] good cooperation in

[46] Council of the European Union, 'Enhanced EU Security Cooperation in and with Asia – Council Conclusions (28 May 2018)', 2018, https://www.consilium.europa.eu/media/35456/st09265-re01-en18.pdf.

[47] Council of the European Union, 3.

[48] South Korea is the EU's eighth largest export destination for goods, whereas the EU is South Korea's third largest export market. See more at: 'Trade with South Korea', European Commission, accessed 22 January 2021, https://ec.europa.eu/trade/policy/countries-and-regions/countries/south-korea/.

achieving legislative coherence,[49] and productive research partnerships,[50] South Korea is an inspiring example of an innovative and high-tech industrialised economy. This was also reflected in the June 2020 virtual meeting of leaders, who were committed to strengthening cooperation in responding to the Fourth Industrial Revolution, harnessing human-centric digitalisation, and promoting the safe and ethical implementation of AI technologies.[51]

A great example of such operationalisation is the pilot project between the EU and South Korea which aims for a deeper and more effective security and defence dialogue, promoting security cooperation and greater convergence between the policies and practices of the EU and South Korea, and increasing international awareness and acknowledgement of the EU as a security provider. Enhanced cooperation on security issues will help to find a balance between the economic and the political/security legs of the EU's cooperation with Asia, and target not only specific Asian countries but also Asian partners outside of the region in this area. In addition to capacity building and awareness raising, the pilot project facilitates agreeing on common approaches at the global level, for example, standards for technologies and norm-building in the area of cybersecurity.[52]

For the purposes of cyber resilience, added value can be seen in activities such as the conduct of peer-to-peer exchanges of technical, legal, and policy experience and best practices, formal and informal dialogues (such as Track 1.5 or 2.0), advising and training, as well as joint exercises. Such activities can be tailored to focus on specific capacities in accordance with relevant international standards and EU best practices needed to face

[49] The European Commission has adopted its adequacy decision for the transfer of personal data from the European Union to the Republic of Korea under the General Data Protection Regulation in 2021. See more at: 'Adequacy Decisions', Text, European Commission, accessed 28 February 2022, https://ec.europa.eu/info/law/law-topic/data-protection/international-dimension-data-protection/adequacy-decisions_en.

[50] European Commission, 'Roadmap for EU – Republic of Korea S&T Cooperation', https://ec.europa.eu/info/sites/info/files/research_and_innovation/strategy_on_research_and_innovation/documents/ec_rtd_eu-korea_roadmap.pdf.

[51] European Commission, 'Joint Press Release: Republic of Korea – EU Leaders' Video Conference Meeting', 30 June 2020, https://ec.europa.eu/commission/presscorner/detail/en/IP_20_1226.

[52] European Commission, 'Partnership Instrument Annual Action Programme 2019, Annex 3 – Action Document for "Security Cooperation in and with Asia"', 2019, 2.

security challenges in the region and beyond, and increase equally the resilience of EU and South Korea.[53]

4 Conclusion

The EU has reiterated its goal to become a global leader in digital affairs on numerous occasions. In order to reach its objective, it must develop and maintain a coherent cybersecurity posture, with relevant legal and policy frameworks harmonising the efforts on member states' level. Investing in, setting standards for, and effectively employing new technologies are a part of building a more resilient EU and boosting EU's defence. In the context of cyber defence, the EU needs to focus on building a security and defence culture, improving the Union's operational effectiveness, resilience, capabilities, and cooperation with partners, as well as defining objectives and concrete goals for its policies. Importantly, technological sovereignty does not entail isolation from other regions' technical solutions but is instead built upon technological innovation and cooperation. Cyber resilience is no longer an issue merely on domestic agendas, but has proved to be a core objective of the EU strategic developments, and also influences the development in the domain of cyber defence. Furthermore, in order to provide stronger protection for the EU against cyber threats, the work towards achieving cyber resilience must reach global dimensions. To that end, international cooperation and strategic partnerships with trusted actors such as South Korea are essential and offer mutual benefits for all.

References

Alessandro Marrone and Ester Sabatino, Cyber Defence in NATO Countries: Comparing Models, Instituto Affari Internazionali, 2021, https://www.iai.it/sites/default/files/iaip2105.pdf.

Anna-Maria Osula, 'The Global Rush for Standards in Blockchain', Directions Blog, 2019, https://directionsblog.eu/the-global-rush-for-standards-in-blockchain/.

Communication from the Commission to the European Parliament, the Council, the European Economic and Social Committee and the Committee of the

[53] Ibid., 5–6.

Regions, Fostering a European approach to Artificial Intelligence, COM(2021) 205 final, 21.04.2021.
Council of the European Union, 'Enhanced EU Security Cooperation in and with Asia – Council Conclusions (28 May 2018)', 2018a.
Council of the European Union, 'EU Cyber Defence Policy Framework (2018 Update) (14413/18)', 2018b.
Elena Lazarou and Alina Dobreva, 'Security and Defence', European Parliamentary Research Service, 2019.
ENISA, 'ENISA Threat Landscape – The Year in Review', 2020, https://www.enisa.europa.eu/publications/year-in-review.
European Commission, 'Adequacy Decisions', https://ec.europa.eu/info/law/law-topic/data-protection/international-dimension-data-protection/adequacy-decisions_en.
European Commission, 'Communication from the Commission to the European Parliament, the Council, the European Economic and Social Committee and the Committee of the Regions – Network and Information Security: Proposal for A European Policy Approach (COM(2001)298)', 2001.
European Commission, 'Communication from the Commission to the European Parliament, the Council, the European Economic and Social Committee and the Committee of the Regions – Cybersecurity Strategy of the European Union: An Open, Safe and Secure Cyberspace (JOIN(2013) 1 Final)', 2013.
European Commission, 'Communication from the Commission to the European Parliament, the Council, the European Economic and Social Committee and the Committee of the Regions on the EU Security Union Strategy (COM/2020/605 Final)', 2020a.
European Union, 'The European Union's Global Strategy: Three Years on, Looking Forward', 2019.
European Commission, 'Joint Communication to the European Parliament and the Council: The EU's Cybersecurity Strategy for the Digital Decade (JOIN (2020)18 Final)', 2020b.
European Commission, 'Joint Press Release: Republic of Korea – EU Leaders' Video Conference Meeting', 30 June 2020c, https://ec.europa.eu/commission/presscorner/detail/en/IP_20_1226.
European Commission, 'Partnership Instrument Annual Action Programme 2019, Annex 3 – Action Document for "Security Cooperation in and with Asia"', 2019a.
European Commission, 'Proposal for a Directive of the European Parliament and of the Council on Measures for a High Common Level of Cybersecurity across the Union, Repealing Directive (EU) 2016/1148 (COM(2020) 823 Final)', 2020d.

European Commission, 'Roadmap for EU – Republic of Korea S&T Cooperation', https://ec.europa.eu/info/sites/info/files/research_and_innovation/strategy_on_research_and_innovation/documents/ec_rtd_eu-korea_roadmap.pdf.
European Commission, 'The Commission Proposes a New Directive to Enhance the Resilience of Critical Entities Providing Essential Services in the EU', Migration and Home Affairs – European Commission, 16 December 2020e, https://ec.europa.eu/home-affairs/news/commission-proposes-new-directive-enhance-resilience-critical-entities-providing-essential_en.
European Commission, 'Trade with South Korea', https://ec.europa.eu/trade/policy/countries-and-regions/countries/south-korea/.
European Commission, "Annexes to the Commission Implementing Decision on the Financing of the European Defence Industrial Development Programme and the Adoption of the Work Programme for the Years 2019 and 2020", C(2019) 2205 final, Brussels, March 19, 2019b, https://ec.europa.eu/docsroom/documents/34515.
European Commission, 2021, Strategic Compass, https://eeas.europa.eu/sites/default/files/towards-a-strategic-compass-2021-february.pdf.
European Commission, European High-Performance Computing Joint Undertaking, https://ec.europa.eu/digital-single-market/en/eurohpc-joint-undertaking.
European Commission, European Defence Industrial Development Programme, 2020 calls for proposals, conditions for the calls and annexe, 2020d.
European Political Strategy Centre, 'Rethinking Strategic Autonomy in the Digital Age', EPSC Strategic Notes, 2019, https://op.europa.eu/en/publication-detail/-/publication/889dd7b7-0cde-11ea-8c1f-01aa75ed71a1/language-en/format-PDF/source-118064052.
European Union Institute for Security Studies, Finding direction with a Strategic Compass? Reflections on the future of EU security and defence, 2020, https://www.iss.europa.eu/sites/default/files/EUISSFiles/Event%20Report%20-%20Strategic%20Compass.pdf.
European Union, 'Proposal for a Directive of the European Parliament and of the Council on the Resilience of Critical Entities (COM(2020) 829 Final)', 2020.
European Union, 'Regulation (EU) 2019/881 of the European Parliament and of the Council of 17 April 2019 on ENISA (the European Union Agency for Cybersecurity) and on Information and Communications Technology Cybersecurity Certification and Repealing Regulation (EU) No 526/2013', 2019.
European Union, 'The Directive (EU) 2016/1148 of the European Parliament and of the Council of 6 July 2016 Concerning Measures for a High Common Level of Security of Network and Information Systems across the Union', 2016.
European Union, A Strategic Compass for Security and Defence, 2022.

Jack Richardson, 'EDIDP Wins for Indra', European Security & Defence, 6 July 2020, https://euro-sd.com/2020/07/articles/18075/edidp-wins-for-indra/.
James Black et al., 'Defence and Security after Brexit: Understanding the Possible Implications of the UK's Decision to Leave the EU – Compendium Report'.
Joint Communication on Resilience, Deterrence and Defence: Building strong cybersecurity for the EU, JOIN (2017) 450.
Luukas Ilves and Anna-Maria Osula, 'The Technological Sovereignty Dilemma – and How New Technology Can Offer a Way Out', European Cybersecurity Journal 6, no. 1 (2020).
PESCO Projects, Cyber Rapid Response Teams and Mutual Assistance in Cyber Security, https://pesco.europa.eu/project/cyber-rapid-response-teams-and-mutual-assistance-in-cyber-security/.
Soare, Simona R, European Defence and AI: Game-Changer or Gradual Change?, 24 March 2021, https://www.rsis.edu.sg/rsis-publication/rsis/european-defence-and-ai-game-changer-or-gradual-change/#.YIPvRqkzY-S.
Ursula von der Leyen, 'Political Guidelines for the Next European Commission 2019–2024', 2019, 13, https://ec.europa.eu/commission/sites/beta-political/files/political-guidelines-next-commission_en.pdf.

Enhancing Cooperation Between South Korea and the EU in the Fight Against Cybercrime

Gibum Kim

G. Kim (✉)
Department of Forensic Science, Sungkyunkwan University, Seoul, South Korea
kgb02@skku.edu

© The Author(s), under exclusive license to Springer Nature Switzerland AG 2022

G. Boulet et al. (eds.), *Cybersecurity Policy in the EU and South Korea from Consultation to Action*, New Security Challenges,
https://doi.org/10.1007/978-3-031-08384-6_10

1 Introduction

Cybercrime is a typical example of transnational crime.[1] Crimes such as hacking, malicious code,[2] child sexual exploitation materials, and ransomware occur across national jurisdictions. While initially considered as a problem of developed countries, cybercrime turned into a global issue with the expansion of the Internet to developing countries since 2000.[3] South Korea was not spared from transnational cybercrime either.[4] When North Korea launched a distributed denial-of-service (DDoS)[5] attack on South Korea and the US in 2009, up to 435 servers in approximately 61 countries were used as intermediate stops.[6] In 2016, the Korean and Australian operators of the Korean porn site Sora.net fled to Malaysia after their services had been interrupted by the police, and transferred their servers from the US to the Netherlands.[7] In 2018, the darkweb website "Welcome to Video" sold child sexual exploitation materials to users in 38 countries receiving Bitcoin.[8] In June 2021, the Korean National Police Agency (KNPA) needed support from the police of the Philippines and

[1] Jonathan Clough, *Principles of Cybercrime(Second Edition)*(New York: Cambridge University Press, 2015), 8.; UNODC, *Comprehensive Study on Cybercrime(Draft)* (New York: United Nations, 2013), 55.

[2] Act on Promotion of Information and Communications Network Utilization and Information, etc. Article 48 (Prohibition on Intrusive Acts, etc. on Information and Communications Network) (2) No one shall mutilate, destroy, alter, or forge an information and communications system, data, program, or similar without a justifiable grounds, nor shall convey or spread a program that is likely to interrupt operation of such system, data, program, or similar (hereinafter referred to as "malicious program").

[3] UNODC, *Comprehensive Study on Cybercrime(Draft)* (New York: United Nations, 2013), 1.

[4] For a cyber threat overview, see the chapter of Mason **Richey**.

[5] Distributed denial-of-service (DDOS) is a type of cybercrime with the aim to prevent a target website from functioning by transmitting multiple connection requests to target servers that exceed the traffic capacity of these servers.

[6] Lee, Seoung Woon, *"A research on the Cyberterrorism response model"* (MS diss., Korea University, 2015), 35.

[7] Kim, Han-kyun et al., *Comprehensive Policy for Developing Scientific Criminal Investigation and Forensic Science (III)* (Seoul: Korea Institute of Criminology, 2020), 289-291.

[8] Korean National Police Agency (KNPA), *White Paper on Policing* (Seoul: KNPA, 2020), 252.

Ukraine to investigate criminal cases where suspects used Gandcrab[9] and Clop[10] ransomware.[11]

Although cybercrime organizations transcend national boundaries, law enforcement agencies (LEAs) are still organized nationally. Their ability to investigate beyond boundaries is also constrained by limits imposed either by domestic or international law: for example, as the international principle of national sovereignty also applies to cyberspace, differences in national legislation hinder cooperation through mutual legal assistance (MLA).[12] In addition, cybercrime organizations use anonymizing technologies such as Tor[13] and Bitcoin,[14] which make it hard for LEAs to collect evidence and trace criminals.

To overcome these issues, South Korea has closely cooperated with foreign LEAs as well as international organizations such as Interpol, Europol, and the United Nations Office on Drugs and Crime (UNODC). The KNPA has hosted an annual conference on cybercrime called "International Symposium on Cybercrime Response (ISCR)" since 2000[15] and signed Memoranda of Understanding (MOU) on countering cybercrime with the US, the UK, France, Germany and the Netherlands.[16] South Korea also provided Interpol with special funds of approximately 8

[9] Gandcrab ransomware is a type of malicious code that encrypts files and demands virtual assets such as Bitcoin. It was discovered in early 2018.

[10] Clop is the name of the ransomware that uses the .clop extension after encrypting files.

[11] Lee, Byung-gil, "Investigation of Ransomware distribution cases", *South Korea digital forensic society*, a presentation at the summer conference, Session 2(June, 2021): 129-142.

[12] Kwon, Han-Young, "International Cooperation for the Prevention of Cyber Terrorism and the Suggestion for Korea", *Dong-A Law Review* 65(2014): 658-659.; Mutual legal assistance is a form of cooperation between different countries for the purpose of collecting and exchanging information.

[13] Tor (The Onion Router) is a web browser for anonymity. It anonymizes the web traffic using the Tor network.

[14] Bitcoin is the first virtual asset. The technology is set out in a white paper by Satoshi Nakamoto in 2008.

[15] Korean National Police Agency (KNPA), *White Paper on Policing* (Seoul: KNPA, 2020), 237.

[16] Korean National Police Agency (KNPA), *White Paper on Policing* (Seoul: KNPA, 2020), 237.

million USD to respond to child sexual exploitation materials and cyber-enabled economic crimes.[17,18]

Despite its seriousness, the level of global cooperation against cybercrime is insufficient. In the case of South Korea, cyber cooperation with the EU and Europol has lagged behind compared to cooperation with the US, Asian countries, and Interpol: the South Korean government is hesitant to join the Convention on Cybercrime of the Council of Europe[19] and does not actively cooperate with European countries, despite several cases of demonstrated successful investigation cooperation such as on the CLOP ransomware cases,[20] and Europol's access to sophisticated investigation capacities in addressing child sexual exploitation, ransomware, and money laundering.[21]

This chapter reviews South Korea's international cooperation framework on cybercrime (Sect. 2) and explains (in Sect. 3) the following four solutions to enhance cooperation between South Korea and the EU: (1) joining the Convention on Cybercrime of the Council of Europe, (2) establishing a working group on joint investigation in cybercrime, (3) implementing cybercrime capacity building projects in developing countries, (4) expanding cooperation on digital forensic technology. Through these measures, South Korea will be able to respond to the global cybercrime threat more actively and take more responsibility for global cybersecurity.

[17] Korean National Police Agency (KNPA), "KNPA funds Interpol to address transnational crimes", KNPA Press release, February 11, 2020; Korean National Police Agency (KNPA), *White Paper on Policing* (Seoul: KNPA, 2020), 276.

[18] Ministry of Culture Sports and Tourism (MCST) Press release (10 Dec 2020).

[19] This Convention is the first international treaty aiming to address legislation related to cybercrime, digital evidence, and international cooperation; Hae-sung Yoon, Ra, Kwang Hyun, "Precedence issues for accession to the Convention on Cybercrime", *Gachon Law Review* 12, no. 3(2019): 203.

[20] Jonathan Greig, "Ukrainian police partner with US, South Korea for raid on Clop ransomware members," *Zdnet*, June 15, 2021.

[21] "Multi-million euro cryptocurrency laundering service Bestmixer.io taken down", Europol Website (May 22, 2019), accessed December 27, 2020, https://www.europol.europa.eu/newsroom/news/multi-million-euro-cryptocurrency-laundering-service-bestmixerio-taken-down.

2 South Korea's International Cooperation Framework on Cybercrime

2.1 *The Change of South Korea's Cybercrime Investigation Authority from the Prosecutor's Office to the KNPA*

Till 2020, most cybercrime investigations in South Korea were carried out by the KNPA and the Prosecutor's Office.[22] In 2020, due to public controversies stemming from the abuse of power of the Prosecutor's Office, the Korean National Assembly revised the Prosecutors' Office Act[23] and drastically limited the Prosecutor's Office investigative mandate to the following categories of serious crimes: (1) specific crimes designated by executive order of the President (Presidential Decree) such as corruption, financial crimes, crimes committed by public officials, election crimes, defence industry crimes, and catastrophic incidents; (2) crimes committed by police officers; (3) crimes identified by the Prosecutor's office with direct relevance for the previous two categories of crimes, and (4) cases sent by the KNPA (Sect. 4 of Prosecutors' Office Act). While preparing the executive order after the enactment of the revised Act, the Prosecutor's Office requested to include cybercrime to the list of "catastrophic incidents" to secure direct authority over cybercrime investigations; however, this request was not accepted by the Presidential Committee on Judicial Reform.

Therefore, the KNPA has become the organization with exclusive investigative authority for cybercrime. It restructured its organization to task cybercrime investigations with the Cyber Investigation Bureau under the National Office of Investigation of the KNPA. As an organization with exclusive jurisdiction over cybercrime investigations, the KNPA is expected to become more effective for promoting international cooperation and implementing effective law enforcement policies on cybercrime.

[22] Digital piracy investigations are carried out by the Special Judicial Police (SJP) of the Ministry of Culture, Sports and Tourism (MCST). The SJP is special investigative authority installed in government departments, specializing in investigating criminal cases in violation of special acts. It undertakes specific functions designated by the Prosecutors' Office supplementing the responsibilities of the national police.

[23] Act No.17566, enforcement date December 8, 2020.

2.2 Progress to Join the Convention on Cybercrime

The Convention on Cybercrime enacted by the Council of Europe covers various crimes committed via the Internet and other computer networks, dealing with violations of network security, computer-related fraud, child sexual exploitation, and infringements of copyright.[24] It also regulates a series of investigation powers and criminal procedures such as the seizure and search of computers and interception.

However, the South Korean government has yet to join the Convention on Cybercrime. South Korea attempted to join the Convention in 2013 during the Global Conference on Cyberspace[25] (led by the UK) but could not make a final decision for several reasons, including (1) the fact that China is not a signatory state (international cooperation with China is critical considering many investigation cases); (2) the challenge of transposing the Convention into domestic procedural law considering the need to broaden investigative power; (3) the numerous already existing (and allegedly effective) channels for international cooperation other than MLA (e.g., Interpol, police-to-police cooperation, LEA-to-Internet Service Providers [ISPs] cooperation, and international cooperation based on the principle of diplomatic reciprocity[26]).

The author believes that these alleged reasons are invalid. First, the Convention on Cybercrime should be seen as an opportunity for South Korea to develop cooperation channels regardless of whether or not China is a signatory state. Second, although updating domestic procedural law (as well as domestic substantive law) may be challenging, such updates are important to meet global standards aimed at a more effective response to cybercrime. Third, there are indeed numerous methods for international cooperation (MLA or otherwise), but they are not sufficient to guarantee binding cooperation. Fortunately, recent cases of investigating hacking, ransomware, and child sexual exploitation convinced the South Korean government that international cooperation is necessary: the President's Office of South Korea announced during the Implementation of the National Cybersecurity Strategy published in 2019 that South Korea

[24] "Convention on Cybercrime (ETS No. 185)", Council of Europe, accessed July 27, 2021, https://www.coe.int/en/web/conventions/full-list?module=treaty-detail&treatynum=185.

[25] The GCCS Conference was proposed by the UK and was held in London. It is a global multistakeholder meeting to promote international cooperation in cyberspace.

[26] These points are the author's summary based on his involvement in the preparation of the Conference on Cyber Space in 2013, and the ROK-US Summit in 2015.

will proceed to join the Convention on Cybercrime.[27] At the time of writing (February 2022) South Korean government authorities including the Ministry of Foreign Affairs, the Ministry of Justice, the Prosecutor's Office, and the KNPA are preparing detailed measures to join the Convention such as revising the Criminal Procedure Law. In addition, academic discussions to join "the Second Additional Protocol to the Convention on Cybercrime on enhanced cooperation and disclosure of electronic evidence (Article 32)" have started.[28]

2.3 Supporting Capacity Building in Global Cybercrime Investigations

Since 2010, the South Korean government has dispatched cybercrime investigation officials abroad and has provided special funding to Interpol to facilitate international investigative cooperation. The first South Korean special funding for an Interpol project was KNPA funding of more than 42 million USD from 2020 to 2022 to support the investigation of child sexual exploitation materials and organized cyber-enabled economic crimes including Voice Phishing.[29] Interpol also established Operation HAECHI[30] under the Cyber Enabled Financial (CEFIN) project to enhance cybercrime capacity building; and the Korean Ministry of Culture, Sports and Tourism (MCST) procured special funds up to 33 million USD from 2021 to 2025 to support the Interpol's counter–digital piracy project named I-SOP (Interpol Stop Online Piracy).[31]

[27] Cheongwadae, National Security Office, *National Cybersecurity Strategy*(Seoul: Cheongwadae, 2019), 1-27.

[28] "Second Additional Protocol to the Convention on Cybercrime on enhanced cooperation and disclosure of electronic evidence by Cybercrime Convention Committee(T-CY)(Ver 12 April 2021)", Council of Europe, accessed Nov 4, 2021, https://rm.coe.int/2nd-additional-protocol-budapest-convention-en/1680a2219c.; Jae Sung Park, "Development of New International Treaties to Combat Cybercrime - Second Additional Protocol to the Budapest Convention and the UN Resolution for a New Cybercrime Treaty -", *The Justice* 185(2021.8): 246-284.

[29] Korean National Police Agency (KNPA), *White Paper on Policing* (Seoul: KNPA, 2020), 276.

[30] Haechi, generally known as Haetae, is a lion-like horned creature that often appears in myths as a guardian against fire and disasters. HAECHI refers to an imaginary animal that judges between right and wrong, and good and evil.

[31] Ministry of Culture, Sports, and Tourism (MCST) press release, December 10, 2020.

In addition, since 2005, South Korea has actively supported investigation and digital forensic capacity building of Asian countries utilizing the Official Development Assistance (ODA) funds of the Korea International Cooperation Agency (KOICA) in close cooperation with the Korea National Police University (KNPU). Best practice examples are two projects in Bangladesh (3 million USD to the National Police HQs from 2012 to 2015, 4.8 million USD to Chattogram Police Agency from 2020 to 2023) and one in Indonesia (5.1 million USD to the National Police HQs from 2018 to 2021). The KNPA also dispatched cybercrime investigators to the Cyber Investigation Bureau of Interpol in Singapore to provide support to training, cooperative investigations, and international meetings.

2.4 International Cooperation with Major Powers Including the US and the EU

The KNPA established a close law enforcement cooperation tie with the FBI when it was investigating hacking cases targeting major Korean government agencies in 2004.[32] In 2006, it also signed a Memorandum of Understanding (MOU) with the FBI to enhance cooperation on cybercrime investigations. This close relationship contributed to the considerable success of the investigation into the massive-scale North Korean DDoS attack of July 7, 2009. Joint investigations and information sharing on cybercrime were intensified in 2013 when the FBI dispatched an agent to the KNPA; and got a further boost in 2015 when the White House made "Stronger cooperation in cybercrime investigation" one of the four cybersecurity agendas in its Joint Statement on Korea-US Relations.[33] As a result, South Korea and the US successfully responded to a series of hacking attempts such as the WannaCry Ransomware 2.0 (presumably

[32] Min Kyung Chung, Jong In Lim, Hun Yeong Kwon, "A Study on North Korea's Cyber Attacks and Countermeasures", *Journal of Information Technology Services* 15, no.1(2016): 70.

[33] The White House (Office of the Press Secretary), "Joint Fact Sheet: The United States-Republic of Korea Alliance: Shared Values, New Frontiers", October 16, 2015.

carried out by North Korea)[34] and apprehended suspects managing the Darkweb website called "Welcome to Video"[35] offering child sexual exploitation materials.

During the ROK-US Summit in June 2021, the two countries agreed to deepen cooperation to ensure an effective joint response against emerging cyber threats.[36] Based on this agreement, the investigation agencies of both countries are working on establishing a cyber working group for ransomware investigations.[37]

Simultaneously, since 2000, the KNPA has maintained a close relationship with European countries such as the UK, Germany, France, and the Netherlands. The DDoS attack against South Korea and the US on July 7, 2009, was resolved through cooperation with Germany,[38] and the success of the SORA.NET case in 2016 was the result of cooperation with the Netherlands.[39]

Europol has achieved great success in developing cybercrime strategies, enhancing digital forensic capacity, and carrying out investigative operations after establishing the European Cybercrime Centre (EC3).[40] It also established a Joint Cybercrime Action Taskforce (J-CAT) in 2014 which consists of a standing operational team of cyber liaison officers from several EU member states and non-EU cooperation partners such as the US and Canada.[41] With the spread of ransomware attacks, South Korea

[34] The White House, "Press Briefing by Press Secretary Sean Spicer and Assistant to the President for Homeland Security and Counterterrorism Tom Bossert", May 5, 2017, accessed June 29, 2021, https://www.presidency.ucsb.edu/documents/press-briefing-press-secretary-sean-spicer-and-assistant-the-president-for-homeland.

[35] U.S. District Court for the District of Columbia Indictment(Case: 1:18-cr-00243), 2018.

[36] "The first ROK-US summit", The President House of ROK, accessed June 29, 2021, https://www1.president.go.kr/articles/10346.

[37] Lee, Byung-gil, "Investigation of Ransomware distribution cases", *South Korea digital forensic society*, a presentation at the summer conference, Session 2 (June, 2021): 129-142.

[38] Lee, Seoung Woon, *"A research on the Cyberterrorism response model"* (MS diss., Korea University, 2015), 35.

[39] Kim, Han-kyun et al., *Comprehensive Policy for Developing Scientific Criminal Investigation and Forensic Science (III)* (Seoul: Korea Institute of Criminology, 2020), 289-291.

[40] European Commission, "The EU Internet Security Strategy in Action: Five steps towards a more secure Europe", Communication from the Commission to the European Parliament and the Council, COM(2010) 673 final, November 21, 2010.

[41] "Joint Cybercrime Action Taskforce (J-CAT)", Europol website, accessed June 29, 2021, https://www.europol.europa.eu/activities-services/services-support/joint-cybercrime-action-taskforce.

recently recognized the need to participate in the J-CAT and is making efforts to join it. Considering Europol's recent success in cybercrime operations such as blocking Bestmixer.io[42](a tool used for laundering cryptocurrencies) and the dismantlement of the Avalanche network[43] (where Phishing and Ransomware groups were engaged), the need for a closer relationship between South Korea and Europol to strengthen cooperation in the fight against cybercrime has become evident.

3 Measures to Enhance International Cybercrime Cooperation: Focus on the EU

3.1 Swiftly Joining the Convention on Cybercrime

South Korea should promptly become a member of the Convention on Cybercrime to respond to transnational cybercrime[44] and surging transnational high-tech cybercrime such ransomware. As a prerequisite, South Korea needs to amend acts containing provisions on criminal procedures such as the Criminal Procedure Act and the Protection of Communication Secrets Act (Act No. 14839, Jul. 26, 2017).

The provisions of the Convention which are not yet codified in South Korean domestic acts include (1) the expedited preservation of stored computer data (article 16),[45] (2) the expedited preservation and partial

[42] "Multi-million euro cryptocurrency laundering service Bestmixer.io taken down", Europol Website (May 22, 2019), accessed December 27, 2020, https://www.europol.europa.eu/newsroom/news/multi-million-euro-cryptocurrency-laundering-service-bestmixerio-taken-down.

[43] "Avalanche Network Dismantled in International Cyber Operation", Europol Website (1 Dec 2016), accessed June 29, 2021, https://www.europol.europa.eu/newsroom/news/'avalanche'-network-dismantled-in-international-cyber-operation.

[44] Lee, Chang Soo, "Measures to enhance international cooperation against the transnational cybercrimes and its prerequisites", *New trends in criminal law* 21(2009.8): 139.; Haesung Yoon, Ra, Kwang Hyun, "Precedence issues for accession to the Convention on Cybercrime", *Gachon Law Review* 12, no. 3(2019): 215.; Chung, Tae-Jin, Rhee, Guang-Meen, "A Study on accession by South Korea to the Budapest convention on cybercrime and international cooperation against cybercrime", *The Police Science Journal* 14, no. 2(2019): 74.

[45] Choi, Hyeok-Doo, "The Review of Implementing Legislation for the Adoption of Convention on Cybercrime", *The Journal of Police Policies* 32, no. 3(2018): 388.

disclosure of traffic data (article 17),[46] and (3) the search and seizure of stored computer data (article 19).[47,48] Codification of these provisions in South Korean law is required as these provisions are not covered by the reservation provision (article 42) that allows signatory states to avail themselves of the reservation(s) provided for in certain provisions of the Convention. In order to codify these provisions, a decision has to be taken which agency—the police, the prosecutor's office, or the court will have the authority to approve expedited preservation. In addition, the public concern stemming from the concentration of investigative authority in one hand needs to be considered to prevent the abuse of power. Though cooperation is reciprocal, the South Korean government has to alleviate the resistance of the public to transfer data to foreign LEAs.[49]

3.2 *Establishing a Working Group on Joint Investigation in Cybercrime Between the EU and South Korea*

The South Korean government needs to establish and actively engage in a joint investigation[50] working group between the EU and South Korea, to address serious cases of cybercrime. The working group needs the participation not only of LEAs but also of the private sector such as ISPs, financial institutions, cryptocurrency exchanges, telecommunication companies, and information security companies. Furthermore, the working group would allow participants to share experiences, to conduct joint investigations, to raise awareness for international cooperation, to empower developing countries (especially Asian countries), and to send a strong message

[46] Park, Hee Young, et al., *A research on the codification of the Convention on Cybercrime* (Seoul: Supreme Prosecutors' Office, 2015), 86.

[47] Park, Hee Young, et al., A research on the codification of the Convention on Cybercrime (Seoul: Supreme Prosecutors' Office, 2015), 111.

[48] One author holds the opinion that Articles 17 and 19 of the Convention can be covered by current Acts (Criminal Procedure Act & Protection of Communications Secrets Act). Choi, Hyeok-Doo, "The Review of Implementing Legislation for the Adoption of Convention on Cybercrime", *The Journal of Police Policies* 32, no. 3(2018): 388-395.

[49] Chung, Tae-Jin, Rhee, Guang-Meen, "A Study on accession by South Korea to the Budapest convention on cybercrime and international cooperation against cybercrime", *The Police Science Journal* 14, no. 2(2019): 73.

[50] A joint investigation happens between countries with jurisdiction over a crime launching investigations in their respective country in accordance with their laws and sharing information with one another. It is an investigation method to overcome the problem of delays and obstacles due to several judicial jurisdictions.

to deter criminals. It would result in more effectively tracking and apprehending suspects, blocking access to illegal websites, banning the distribution of illegal contents, suspending payments from financial accounts, and tracing money laundering. The mutual exchange of cybercrime investigators or digital forensic experts is a potential means to facilitate joint investigations. The achievements of the working group would also be a catalyser to joining the Convention on Cybercrime.

3.3 Delivering Cybercrime Capacity Building Projects in Developing Countries

South Korea and the EU could jointly enhance cybercrime investigation capabilities of developing countries. The EU has significant experience in international development cooperation targeting LEAs of developing countries. If South Korea and the EU can jointly implement capacity building projects, also in terms of logistics, there will be strong synergies, avoiding or at least reducing the duplication of efforts in delivering trainings under South Korean KOICA projects and the EU GLACY+ project.[51] Furthermore, both will be able to jointly address emerging cybercrime trends in developing countries; leading joint investigations; delivering trainings on digital forensic techniques; identifying best practices; supporting operations on ransomware, darkweb, and child sexual exploitation; and taking measures to prevent the infringement of intellectual property rights.

3.4 Expanding Cooperation on Digital Forensic Technology

In addition to cooperation on cybercrime investigations, South Korea and the EU also need to promote cooperation on digital forensic[52] technology (including research and information sharing on technologies such as malicious code and Bitcoin analysis) as well as on mobile forensics. This will contribute to standardizing digital forensic procedures facilitating the exchange of digital evidence between countries. In addition, there will be

[51] "Global Action on Cybercrime Extended (GLACY)+", INTERPOL Website, accessed June 29, 2021, https://www.interpol.int/Crimes/Cybercrime/Cyber-capabilities-development/Glacy.

[52] Interpol defines digital forensics as a branch of forensic science that focuses on identifying, acquiring, processing, analysing, and reporting on data stored electronically. See https://www.interpol.int/en/How-we-work/Innovation/Digital-forensics

positive spill-over effects to address a wide variety of new technology issues such as autonomous vehicles, deep-fake, and crimes utilizing artificial intelligence (AI).

In addition to the support by LEAs, there should also be buy-in by universities/research institutes, standardization organizations, and private companies. Good examples of such cooperative actions are the Global Cybercrime Expert Group operated by Interpol since 2016, and a research paper on AI-related crime by Trend Micro Research, the United Nations Interregional Crime and Justice Research Institute (UNICRI), and Europol's European Cybercrime Centre (EC3).[53]

3.5 *Signing an Executive Agreement of the CLOUD Act*

The South Korean government should positively consider signing an "executive agreement" as stipulated in the US Clarifying Lawful Overseas Use of Data Act (CLOUD Act).[54] This act, which partially replaces mutual legal assistance in criminal matters,[55]results from the court case *the United States v. Microsoft Corporation*.[56] As a result, the US Congress enacted the CLOUD Act in 2018, which authorizes US LEAs to issue warrants to US ISPs to obtain electronic communications stored abroad and allows US ISPs to disclose US-stored data to foreign LEAs after entering into an executive agreement with the US government.[57] The UK and Australian governments made use of the possibility of an executive agreement, as

[53]Trend Micro Research, United Nations Interregional Crime and Justice Research Institute (UNICRI), Europol's European Cybercrime Centre (EC3), "Malicious Uses and Abuses of Artificial Intelligence", 2020.

[54]See also the chapter of Tatiana **Tropina**.

[55]Song, Young-jin, "Passage of the CLOUD Act and Its Implications for Law Enforcement Access to Extraterritorial Data", *Korean Criminological Review* 29, no 2(2018): 163-164.; Hae-sung Yoon, Ra, Kwang Hyun, "Precedence issues for accession to the Convention on Cybercrime", *Gachon Law Review* 12, no. 3(2019): 215.; Kim, Jae-Woon, "Review on the Counter-Terrorism Policy about CLOUD Act in U.S.A", *Korean Terrorism Studies Review* 11, no 3(2018): 160.

[56]The United States v. Microsoft Corp., 138 S. Ct. 1186, 200 L. Ed. 2d 610 (2018).

[57]Jeon, Chi-Hon, "Extra-Territorial Seizure of Digital Evidence in USA", *Theories and Practices of Criminal Procedure* 10, no 2(2018): 96-104.; Hae-sung Yoon, Ra, Kwang Hyun, "Recedence issues for accession to the Convention on Cybercrime", *Gachon Law Review* 12, no. 3(2019): 214.

approved by the "Crime (Overseas Production Orders) Act 2019" respectively the "Telecommunications Legislation Amendment (International Production Orders) Bill 2020" of 2021.[58,59]

4 Conclusion

This chapter reviewed South Korea's international cooperation on cybercrime and presented four constructive proposals to enhance international cooperation between South Korea and the EU.

First, South Korea should promptly join the Convention on Cybercrime of the Council of Europe. Second, South Korea should make efforts to establish a cybercrime joint investigation working group with the EU, with the participation of LEAs, international organizations, and private companies. Third, South Korea and the EU should strengthen international cooperation in delivering cybercrime capacity building for developing countries. As both South Korean and EU projects would benefit from less (or no) duplication of efforts, collective efforts could be taken such as investigating cybercrime in developing countries, conducting joint investigations, delivering trainings on digital forensic techniques, identifying best practices, and supporting standard-setting operations. Lastly, both parties need to expand cooperation on digital forensic technology, which is critical in cybercrime investigations, engage in active discussions of emerging cyber threats, and develop response strategies making use of developing crime technologies. Through these efforts, South Korea will be able to shoulder more responsibilities and play a greater role on the international stage.

[58] Jennifer Daskal, "Microsoft Ireland, the CLOUD Act, and International Lawmaking 2.0", *Stanford Law Review Online* 71(2018): 14.; "Crime(Overseas Production Orders) Act 2019",

Legislation Website of the UK, accessed June 29, 2021, https://www.legislation.gov.uk/ukpga/2019/5/contents/enacted.

[59] "Telecommunications Legislation Amendment (International Production Orders) Bill 2020", Parliament of Australia, access June 29, 2021, https://www.aph.gov.au/Parliamentary_Business/Bills_LEGislation/Bills_Search_Results/Result?bId=r6511.

References

Cheongwadae, National Security Office. *National Cybersecurity Strategy.* Seoul: Cheongwadae, 2019.

Choi, Hyeok-Doo, "The Review of Implementing Legislation for the Adoption of Convention on Cybercrime", *The Journal of Police Policies* 32, no. 3(2018): 388.

Chung, Tae-Jin, Rhee, Guang-Meen, "A Study on accession by South Korea to the Budapest convention on cybercrime and international cooperation against cybercrime", *The Police Science Journal* 14, no. 2(2019): 74.

Council of Europe. "Convention on Cybercrime (ETS No. 185)." Accessed July 27, 2021a. https://www.coe.int/en/web/conventions/full-list?module=treaty-detail&treatynum=185.

Council of Europe. "Second Additional Protocol to the Convention on Cybercrime on enhanced cooperation and disclosure of electronic evidence by Cybercrime Convention Committee(T-CY)(Ver 12 April 2021b)." Accessed November 4, 2021. https://rm.coe.int/2nd-additional-protocol-budapest-convention-en/1680a2219c.

European Commission, "The EU Internet Security Strategy in Action: Five steps towards a more secure Europe", Communication from the Commission to the European Parliament and the Council, COM(2010) 673 final, November 21, 2010.

Europol Website. "Multi-million euro cryptocurrency laundering service Bestmixer.io taken down", Accessed December 27, 2020. https://www.europol.europa.eu/newsroom/news/multi-million-euro-cryptocurrency-laundering-service-bestmixerio-taken-down.

Kim, Han-kyun et al. *Comprehensive Policy for Developing Scientific Criminal Investigation and Forensic Science(III).* Seoul: Korea Institute of Criminology, 2020.

Kim, Jae-Woon, "Review on the Counter-Terrorism Policy about CLOUD Act in U.S.A", *Korean Terrorism Studies Review* 11, no 3 (2018): 160.

Europol website. "Joint Cybercrime Action Taskforce (J-CAT)" Accessed June 29, 2021a. https://www.europol.europa.eu/activities-services/services-support/joint-cybercrime-action-taskforce.

Europol Website. "Avalanche Network Dismantled in International Cyber Operation" Accessed June 29, 2021b, https://www.europol.europa.eu/newsroom/news/'avalanche'-network-dismantled-in-international-cyber-operation.

Hae-sung Yoon, Ra, Kwang Hyun, "Recedence issues for accession to the Convention on Cybercrime", *Gachon Law Review* 12, no. 3(2019): 203.

INTERPOL Website. "Global Action on Cybercrime Extended (GLACY)+." Accessed June 29, 2021. https://www.interpol.int/Crimes/Cybercrime/Cyber-capabilities-development/Glacy.

Jae Sung Park, "Development of New International Treaties to Combat Cybercrime - Second Additional Protocol to the Budapest Convention and the UN Resolution for a New Cybercrime Treaty -", *The Justice* 185(2021.8): 246-284.

Jennifer Daskal, "Microsoft Ireland, the CLOUD Act, and International Lawmaking 2.0", Stanford Law Review Online 71 (2018): 14.

Jeon, Chi-Hon, "Extra-Territorial Seizure of Digital Evidence in USA", *Theories and Practices of Criminal Procedure* 10, no 2 (2018): 96-104.

Jonathan Clough. *Principles of Cybercrime (Second Edition)*. New York: Cambridge University Press, 2015.

Jonathan Greig. "Ukrainian police partner with US, South Korea for raid on Clop ransomware members." *Zdnet*, June 15, 2021.

Korean National Police Agency (KNPA). *White Paper on Policing*. Seoul: KNPA, 2020a.

Korean National Police Agency (KNPA). "KNPA funds Interpol to address transnational crimes". KNPA Press release, February 11, 2020b.

Kwon, Han-Young, "International Cooperation for the Prevention of Cyber Terrorism and the Suggestion for Korea", *Dong-A Law Review* 65(2014): 658-659.

Lee, Byung-gil, "Investigation of Ransomware distribution cases", *South Korea digital forensic society*, a presentation at the summer conference, Session 2 (June, 2021): 129-142.

Lee, Chang Soo, "Measures to enhance international cooperation against the transnational cybercrimes and its prerequisites", *New trends in criminal law* 21 (2009.8): 139.

Lee, Seoung Woon. *"A research on the Cyberterrorism response model."* MS diss., Korea University, 2015.

Legislation Website of the UK. "Crime(Overseas Production Orders) Act 2019." Accessed June 29, 2021. https://www.legislation.gov.uk/ukpga/2019/5/contents/enacted.

Min Kyung Chung, Jong In Lim, Hun Yeong Kwon, "A Study on North Korea's Cyber Attacks and Countermeasures", *Journal of Information Technology Services* 15, no. 1 (2016): 70.

Park, Hee Young, et al. *A research on the codification of the Convention on Cybercrime*. Seoul: Supreme Prosecutors' Office, 2015.

Parliament of Australia. "Telecommunications Legislation Amendment (International Production Orders) Bill 2020." Accessed June 29, 2021. https://www.aph.gov.au/Parliamentary_Business/Bills_LEGislation/Bills_Search_Results/Result?bId=r6511.

Song, Young-jin, "Passage of the CLOUD Act and Its Implications for Law Enforcement Access to Extraterritorial Data", *Korean Criminological Review* 29, no 2 (2018): 163-164.

The President House of ROK. "The first ROK-US summit." Accessed June 29, 2021. https://www1.president.go.kr/articles/10346.

The White House (Office of the Press Secretary). "Joint Fact Sheet: The United States-Republic of Korea Alliance: Shared Values, New Frontiers", October 16, 2015.

The White House in the U.S.A, "Press Briefing by Press Secretary Sean Spicer and Assistant to the President for Homeland Security and Counterterrorism Tom Bossert", May 5, 2017, accessed 29 June, 2021, https://www.presidency.ucsb.edu/documents/press-briefing-press-secretary-sean-spicer-and-assistant-the-president-for-homeland.

Trend Micro Research, United Nations Interregional Crime and Justice Research Institute (UNICRI), Europol's European Cybercrime Centre (EC3), "Malicious Uses and Abuses of Artificial Intelligence", 2020.

UNODC. *Comprehensive Study on Cybercrime(Draft)*. New York: United Nations, 2013.

The EU and Access to Electronic Evidence: Privatisation of Law Enforcement?

Tatiana Tropina

1 Introduction[1]

Debates about cross-border access to electronic evidence in recent years have gone far beyond the issues related to cybercrime investigations and have taken a central stage in the criminal justice policymaking processes in the EU and around the globe.[2] Nowadays, any crime can leave digital traces. There is a growing demand from law enforcement to get quick access to highly volatile data not only in cases where the victim and

[1] The author would like to sincerely thank Gertjan **Boulet** and Michael **Reiterer** for their valuable feedback on the draft of this chapter and their tireless efforts to make this book a success.

[2] Sergio Carrera, Marco Stefan and Valsamis Mitsilegas, "Cross-border data access in criminal proceedings and the future of digital justice", (2020), 1, https://www.ceps.eu/wp-content/uploads/2020/10/TFR-Cross-Border-Data-Access.pdf

T. Tropina (✉)
Leiden University, Leiden, Netherlands
e-mail: t.tropina@fgga.leidenuniv.nl

© The Author(s), under exclusive license to Springer Nature Switzerland AG 2022

G. Boulet et al. (eds.), *Cybersecurity Policy in the EU and South Korea from Consultation to Action*, New Security Challenges, https://doi.org/10.1007/978-3-031-08384-6_11

offender are in different countries but also in purely domestic cases when it is only data that is stored with a communications provider located in a different jurisdiction.[3] The importance of improving international cooperation to secure cross-border access to data has become a truism. Yet various policy initiatives, rather than focusing on enhancing current mechanisms of cooperation between competent authorities, aim to enable gathering e-evidence via mandatory direct requests to providers located in a foreign jurisdiction.

Unilaterally claiming extraterritorial jurisdiction and forcing foreign providers to cooperate upon requests from abroad might seem like an attractive solution to fix a slow traditional system of cooperation between competent authorities. This solution, however, ignores the crucial reasons for the existence of mutual legal assistance in the first place. By making foreign requests the subject of scrutiny of the competent authorities in the executing country, the mechanism of mutual legal assistance provides an additional layer of protection of fundamental rights.[4] This protection is the primary reason why cross-border access to evidence is governed in accordance with the law of both states[5] with the imposition of additional safeguards, such as dual criminality requirements, lists of offences that are subject to investigative measures, and protection of certain types of privileged communications such as lawyers, medical professions, and others.

The EU places commitment to human rights and the rule of law at the core of its fundamental values.[6] In light of this commitment, the EU could be expected to champion human rights in international negotiations on frameworks related to cross-border digital investigations. However, caught between the growing demand for quick access to data and the need to adhere to its values, the EU is still trying to find its stand in the current debates about improving mechanisms to obtain e-evidence.

[3] European Commission, "Commission staff working document. Impact assessment", April 17, 2018, https://eur-lex.europa.eu/legal-content/EN/TXT/PDF/?uri=CELEX:52018SC0118&from=EN

[4] European Parliament, "An assessment of the Commission's proposals on electronic evidence", (2018), 21-22, https://www.europarl.europa.eu/RegData/etudes/STUD/2018/604989/IPOL_STU(2018)604989_EN.pdf; Els De Busser, "The Digital Unfitness of Mutual Legal Assistance," *Security and Human Rights*, no. 1 (2017).

[5] Carrera, Stefan and Mitsilegas, ii

[6] See: European Union, "EU in Brief. Values", https://europa.eu/european-union/about-eu/eu-in-brief_en

This contribution analyses the recent developments in cross-border access to e-evidence and related challenges for the EU by focussing on three instruments: the US CLOUD Act 2018, the EU E-Evidence Package (hereinafter also referred to as E-evidence proposals), and the Second Additional Protocol to the Convention on Cybercrime (hereinafter Budapest Convention).[7] Section 2 of this chapter starts the discussion by placing the EU mechanisms for access to e-evidence in the context of global challenges. The chapter proceeds in Sections 3, 4, and 5, with analysing the implications of three instruments that aim to improve cross-border access to e-evidence: the US CLOUD Act, the EU E-evidence proposals, and the Second Additional Protocol to Budapest Convention. Ultimately, based on the drawbacks of the presently debated solutions, Section 6 of this chapter argues that there is a need to employ more scrutiny for the mechanisms based on direct requests to providers and, ultimately, to improve existing mutual legal assistance frameworks rather than circumvent them.

2 The EU Approach to E-evidence: Mechanisms and Challenges

In the recent decade, the EU, based on the mutual trust between its member states, has developed a system for cooperation in obtaining evidence across its internal borders. This cooperation is based on the mutual recognition of judicial decisions and on the possibility for judicial authorities to communicate directly. The cross-border cooperation on obtaining electronic evidence between the EU member states relies on the mechanisms

[7] US Cloud Act, https://www.justice.gov/dag/page/file/1152896/download; European Commission, "Proposal for a Directive of the European Parliament and of the Council laying down harmonised rules on the appointment of legal representatives for the purpose of gathering evidence in criminal proceedings", (2017), https://eur-lex.europa.eu/legal-content/EN/TXT/?uri=COM%3A2018%3A226%3AFIN , European Commission, "Proposal for a Regulation of the European Parliament and of the Council on European Production and Preservation Orders for electronic evidence in criminal matters", 2018, https://eur-lex.europa.eu/legal-content/EN/TXT/?uri=COM%3A2018%3A225%3AFIN ; Cybercrime Convention Committee (T-CY), "(Draft) Second Additional Protocol to the Convention on Cybercrime on enhanced co-operation and disclosure of electronic evidence as approved by the T-CY at its 24th Plenary", May, 28, 2021, https://rm.coe.int/0900001680a42c4b

provided in the Directive[8] on European Investigation Order (EIO),[9] which establishes standards for requests and deadlines for recognition and execution that aim to facilitate information exchange in a timely manner. International cooperation with the countries outside the EU is governed by various bilateral and multilateral treaties. For certain types of crimes, this includes international treaties such as the UN Convention on Transnational Organised Crime and the Budapest Convention. Bilateral agreements exist both between individual member states—the EU members and non-EU members, but notably also between the European Union (as a bloc) and non-EU countries, such as the agreement on mutual legal assistance between the EU and the US.[10] There is no agreement between the EU as a block and South Korea: Article 31.3 of the EU-ROK Framework Agreement lays out the commitment to enhance agreements on mutual legal assistance on criminal matters, including, where appropriate, accession to and cooperation with the instruments of the United Nations.[11]

The current agreements, both within the EU and with the countries outside the bloc, cannot solve the issues related to access to electronic evidence. The mutual legal assistance mechanisms have been criticised as slow[12] and digitally "unfit",[13] which is not surprising as they were designed

[8] As this chapter refers to the EU Directives and Regulations, it is important to highlight the difference between these legislative acts. The EU Directives set the goals that must be achieved by the member states, but each member state decides how to transpose them into their national laws. The EU Regulations are binding acts and apply in their entirety to each of the EU member state.

[9] European Union, "Directive 2014/41/EU of the European Parliament and of the Council of 3 April 2014 regarding the European Investigation Order in criminal matters", (2014), https://eur-lex.europa.eu/legal-content/EN/TXT/?uri=celex%3A32014L0041

[10] See: European Union, "Agreement with the United States on mutual legal assistance", July 19, 2003, https://eur-lex.europa.eu/legal-content/EN/TXT/?uri=LEGISSUM%3Ajl0052

[11] European Union, "Framework Agreement between the European Union and its Member States, on the one part, and the Republic of Korea, on the other part", January 23, 2013, https://eur-lex.europa.eu/legal-content/EN/TXT/?uri=CELEX%3A22013A0123%2801%29

[12] See e.g. European Commission, "Commission staff working document. Impact assessment", April 17, 2018; Gail Kent, "The Mutual Legal Assistance Problem Explained", (2015), http://cyberlaw.stanford.edu/blog/2015/02/ mutual-legal-assistance-problem-explained; Andrew K. Woods, "Data Beyond Borders: Mutual Legal Assistance in the Internet Era", January 27, 2015, https://uknowledge.uky.edu/law_facpub/518/

[13] De Busser, The Digital Unfitness.

well before the era of digital investigations. Even the shortened timelines for addressing requests, such as provided in the European Investigation Order (30 days for recognition, 90 days for answering it after the recognition decision[14]), cannot solve the issue of the volatility data and do not meet the growing demand to obtain the data promptly. The EIO Directive also allows requesting "provisional measures"[15] that have to be applied within 24-hours of receiving the request to prevent the destruction of information. However, while these measures can be used to "freeze" the data before the competent authorities send the order to obtain it, the overall timeline still does not help to get information quickly.[16] As to cooperation with non-EU countries, the waiting times are much longer, even with signatories of the Budapest Convention. For example, on average, it takes ten months for the US authorities to answer requests coming from abroad.[17]

The situation is getting even more complicated due to the existence of voluntary channels for direct cooperation with communication providers to obtain certain types of data.[18] For example, US providers can respond to direct requests from foreign law enforcement for traffic and subscriber data, even if they are not obliged to do so by law.[19] However, the content data can only be provided upon mutual legal assistance request and authorisation of a US court.[20] While some of the US providers have developed their own practices on addressing the requests for non-content data

[14] European Union, Directive 2014/41/EU, Art. 12

[15] European Union, Directive 2014/41/EU, Art.32

[16] European Commission, Commission staff working document

[17] Jennifer Daskal. "A New UK-US Data Sharing Agreement: A Tremendous Opportunity, If Done Right", (2016), https://www.justsecurity.org/29203/british-searches-america-tremendous-opportunity/

[18] Stanislaw Tosza, "Cross-Border Gathering of Electronic Evidence : Mutual Legal Assistance, Its Shortcomings and Remedies", *Société numérique et droit pénal*, edited by Daniel Flore and Vanessa Franssen, (Bruylant: 2019), 269

[19] Mirko Hohmann and Sophie Barnett. "System Upgrade. Improving Cross-Border Access to Electronic Evidence", January 21, 2019, https://www.gppi.net/2019/01/21/system-upgrade

[20] Marco Stefan and Gloria González Fuster, "Cross-border Access to Electronic Data through Judicial Cooperation in Criminal Matters State of the art and latest developments in the EU and the US", December 3, 2018, https://www.ceps.eu/ceps-publications/cross-border-access-electronic-data-through-judicial-cooperation-criminal-matters/

from foreign law enforcement,[21] not all of them cooperate voluntarily. This patchwork of approaches creates uncertainty for law enforcement agencies in the EU countries who have become dependent on the willingness of private parties to answer requests.[22]

The European Commission estimated, based on a survey of public authorities, that less than half of requests sent abroad—both via mutual legal assistance channels and directly to the service providers—are being answered.[23] While this data can raise the question of possible bias and flaws in the survey as a method to collect statistics on the matter,[24] the very existence of these estimates reflects the growing frustration with the current mechanisms for obtaining e-evidence. There are, of course, a variety of possible reasons for the request not being answered, which does not always involve providers' reasons for refusal to cooperate. For example, requests might be incomplete or seek data providers do not have. Yet undeniably, the absence of unified practices on the side of both law enforcement and providers further contributes to uncertainties.

Another factor that increases uncertainty is that the growing number of EU member states started adopting unilateral approaches to obtaining evidence stored abroad without resorting to mutual legal assistance mechanisms. These approaches, among other cases, also concern crime investigations when both a victim and a criminal are located on the member state's soil, and the only link to the foreign country is that electronic evidence is stored with a provider abroad. Notable cases of unilateral approaches include the Yahoo! case in Belgium when Yahoo! was convicted for not complying with the domestic request from the Belgian prosecutor to provide subscriber data; and the similar Skype case where Skype was convicted for refusing to cooperate with Belgian law enforcement

[21] Piotr Bąkowski and Sofija Voronova, "Electronic evidence in criminal matters. Briefing. EU Legislation in Progress", (March 2021), 3, https://www.europarl.europa.eu/RegData/etudes/BRIE/2021/690522/EPRS_BRI(2021)690522_EN.pdf; see also SIRIUS Project, "EU Digital Evidence Situation Report. 2nd Annual Report", December 1, 2020, 22, https://www.europol.europa.eu/publications-documents/sirius-eu-digital-evidence-situation-report-2nd-annual-report

[22] SIRIUS Project, 15

[23] European Commission, Commission staff working document, 15-16.

[24] Sergi Vazquez Maymir. "Anchoring the need to revise cross-border access to e-evidence", *Internet Policy Review 9 (3), (2020)*, 7.

authorities in obtaining content data.[25] Both cases led to the incorporation of broad provisions on providers' duty to cooperate in the Belgian Code of Criminal Procedure.[26] The summary of responses to the European Commission questionnaire indicated that several EU member states—among them, Belgium, Cyprus, Spain, Portugal, France, and Lithuania—consider requests from their national law enforcement agencies sent to providers abroad without using mutual legal assistance mechanisms as mandatory.[27] Yet despite their perceived compulsory nature, the enforceability of these requests is questionable[28] because there are no clear frameworks to force foreign providers, which are out of the reach of domestic law enforcement, to hand over e-evidence.

This patchwork of national legislative approaches and individual providers' policies on voluntary cooperation creates the situation when neither law enforcement nor providers can be sure which data can be obtained and under which regime: voluntary cooperation, a mandatory domestic request from law enforcement authorities, or a mutual legal assistance request. In an attempt to solve this problem, several legislative solutions emerged in recent years, notably: the US CLOUD (The Clarifying Lawful Overseas Use of Data) Act 2018, the EU E-Evidence proposals 2018, and the Second Additional Protocol to the Budapest Convention. To a greater or lesser extent, these frameworks aim to replace the current mutual legal assistance mechanisms with direct requests to providers located abroad.

3 US CLOUD Act

The first solution of this kind—the US CLOUD Act—was adopted as a response to the *Microsoft Ireland* case, where the US company Microsoft had to provide data upon a warrant issued in the US.[29] In response to the

[25] Paul de Hert, Cihan Parlar and Johannes Thumfart. "Legal Arguments Used in Courts Regarding Territoriality and Cross-Border Production Orders: From Yahoo Belgium to Microsoft Ireland", *New Journal of European Criminal Law 9, no. 3 (September 2018)*, 18-20.

[26] SIRIUS Project, 16

[27] European Commission, "Questionnaire on improving criminal justice in cyberspace Summary of Responses", (n/d), 2, https://ec.europa.eu/home-affairs/sites/default/files/e-library/documents/policies/organized-crime-and-human-trafficking/e-evidence/docs/summary_of_replies_to_e-evidence_questionnaire_en.pdf

[28] Ibid

[29] Jennifer Daskal, "Microsoft Ireland, the CLOUD Act, and International Lawmaking 2.0", *Stanford Law Review Online, Vol. 71, (2018).*

request, Microsoft handed over subscriber data and other non-content data but refused to disclose content data stored abroad on the servers in Ireland. Microsoft argued that the US authorities had to seek data from Ireland under the mutual legal assistance procedure.[30] The case was under consideration in the US Supreme Court when the CLOUD Act was adopted and was subsequently mooted.[31]

The US CLOUD Act obliges all the providers on the US territory to disclose the data, particularly content data, independent of the location where the data is stored. In terms of requests coming from abroad, the Act enables the US to enter executive agreements with foreign governments to allow for requests for data to be sent directly to the service providers in the US.[32] These executive agreements, however, concern data of foreign citizens and residents and do not apply to US residents and citizens (both natural and legal persons).

The CLOUD Act has been widely criticised for unilaterally circumventing safeguards existing in foreign jurisdictions by obliging US companies to provide data stored abroad.[33] In relation to the EU, there are also specific concerns about the GDPR and requirements that must be fulfilled for transferring the personal data of EU citizens. The service providers are in an uncertain position. On the one hand, they must follow the requirements set by the CLOUD Act and provide the data upon request of the US authorities. On the other hand, they face risks of violating the EU laws and imposition of huge fines,[34] especially considering the European Data Protection Board (hereinafter EDPB) Guideline on the Article 49 GDPR, obliging the EU companies to "refuse direct requests and refer the

[30] Hert, Parlar and Thumfart, 20

[31] See: United States v. Microsoft Corp., 584 US ___ (2018), https://supreme.justia.com/cases/federal/us/584/17-2/

[32] US Cloud Act

[33] Katitza Rodriguez, "The US CLOUD Act and the EU: A Privacy Protection Race to the Bottom", April 9, 2018, https://www.eff.org/fr/deeplinks/2018/04/us-cloud-act-and-eu-privacy-protection-race-bottom ; Lawrence Siry, "Cloudy Days Ahead: Cross-Border Evidence Collection and Its Impact on the Rights of EU Citizens", *New Journal of European Criminal Law 10, no. 3 (September 2019):* 227–50.

[34] Emmanuelle Mignon, "The CLOUD Act : Unveiling European Powerlessness", September5,2020,https://legrandcontinent.eu/fr/2020/09/05/the-cloud-act-unveiling-european-powerlessness/ ; Marcin Rojszczak, "CLOUD act agreements from an EU perspective", *Computer Law & Security Review, Volume 38, (2020)*.

requesting third country authority to an existing mutual legal assistance treaty or agreement".[35]

The European Commission started negotiations with the US on the agreement under the CLOUD Act in 2019.[36] Even before the start of negotiations, the Commission acknowledged that the agreement could affect the fundamental rights of individuals and the rights of service providers.[37] Furthermore, the EDPB voiced GDPR-related concerns and called for appropriate safeguards to comply with EU data protection law.[38] The negotiation process further highlighted fundamental problems such as differences between the US and EU laws in terms of definitions related to e-evidence, the types of offences that enable the exchange of e-evidence, and, more importantly, the role of judicial authorities in issuing requests for data.[39] Even the nature of the possible agreement is problematic. The US CLOUD Act envisages bilateral agreements with national governments. Therefore, the US supports the idea of a general agreement with the EU followed by bilateral agreements with individual member states. The European Commission, however, insists on an agreement as a bloc to guarantee that member states are treated equally and to avoid fragmentation.[40] Furthermore, the EU is currently negotiating new mechanisms to establish a procedure for obtaining data from foreign providers. As the

[35] EDPB, "Guidelines 2/2018 on derogations of Article 49 under Regulation 2016/679", (2018), 5, https://edpb.europa.eu/ourwork-tools/our-documents/guidelines/guidelines-22018-derogations-article-49-under-regulation_en

[36] European Commission, "Criminal justice: Joint statement on the launch of EU-US negotiations to facilitate access to electronic evidence", September, 26, 2019, https://ec.europa.eu/commission/presscorner/detail/en/STATEMENT_19_5890

[37] European Commission, "Recommendation for a Council Decision authorising the opening of negotiations in view of an agreement between the European Union and the United States of America on cross-border access to electronic evidence for judicial cooperation in criminal matters", February 5, 2019, https://eur-lex.europa.eu/legal-content/EN/TXT/?uri=CELEX%3A52019PC0070

[38] EDPB, "Annex. Initial legal assessment of the impact of the US CLOUD Act on the EU legal framework for the protection of personal data and the negotiations of an EU-US Agreement on cross-border access to electronic evidence", (2019), 10, https://edpb.europa.eu/sites/default/files/files/file2/edpb_edps_joint_response_us_cloudact_annex.pdf

[39] Thomas Wahl, "E-Evidence: Start of Negotiations on EU-US Agreement", January 12, 2020, https://eucrim.eu/news/e-evidence-start-negotiations-eu-us-agreement/

[40] Theodore Christakis and Fabien Terpan, "EU-US Negotiations on Law Enforcement Access to Data: Divergences, Challenges and EU Law Procedures and Options", *International Data Privacy Law, 11 (2), (2021),* 1

European Commission wants the agreement on the US CLOUD Act to be compatible with the future EU e-evidence mechanisms,[41] the CLOUD Act negotiations depend on the outcome of the current debate about frameworks for e-evidence in the EU.

4 EU E-Evidence Proposals

The EU E-evidence proposals of 2018 consist of two legislative frameworks: a Regulation on European production and preservation orders[42] and a Directive on legal representatives in the EU member states.[43] The Regulation enables the competent authorities in one EU member state to send a direct request to another EU member state to either get access to e-evidence (production order) or preserve specific e-evidence (preservation order). The Directive proposal obliges service providers operating in one or more EU member states to appoint at least one legal representative in the EU to answer requests for e-evidence.

The E-evidence proposals were met with criticism and disapproval by various stakeholders—civil society organisations, communications providers and digital industry, academia, and public authorities in some EU countries. The crux of replacing the mutual legal assistance mechanisms with direct requests is the question of who will be responsible for the protection of fundamental rights. By removing the involvement of competent authorities in the requested states in evaluating requests for data, the E-evidence proposals are delegating the decisions about assessing the validity and legality of the requests from abroad to private entities. However, are the providers in a position to make such an assessment and decide whether to grant a request or reject it? The possibility to refuse a request is crucial for protecting fundamental rights, and so is the process of scrutinising requests by competent authorities. It has been a growing concern that the E-evidence proposals take a step to privatise law

[41] European Commission, "Recommendation for a Council Decision".

[42] European Commission, "Proposal for a Regulation of the European Parliament and of the Council on European Production and Preservation Orders for electronic evidence in criminal matters", April 17, 2018, https://eur-lex.europa.eu/legal-content/EN/TXT/?uri=COM%3A2018%3A225%3AFIN

[43] European Commission, "Proposal for a Directive of the European Parliament and of the Council laying down harmonised rules on the appointment of legal representatives for the purpose of gathering evidence in criminal proceedings", April 17, 2018, https://eur-lex.europa.eu/legal-content/EN/TXT/?uri=COM%3A2018%3A226%3AFIN

enforcement by putting responsibility for protecting fundamental rights on private entities, which are neither legitimate nor competent authorities to bear this responsibility.[44] By choosing this approach, the proposals turn the communication providers into de facto "judicial authorities"[45] and may "lead to a race to the bottom in terms of fundamental rights protection at a global level".[46]

The E-evidence package is currently under negotiation. Shortly before the start of the negotiations, in December 2020, the European Parliament's LIBE Committee (Committee on Civil Liberties) adopted a set of compromise proposals,[47] making several suggestions for improvements. The main proposed revision includes a notification mechanism for the state where a provider is located (executing state) and the member state of the permanent residence of the person affected by request. Under the proposed mechanism, the executing state would have a right to object to the request based on the protection of human rights. This notification procedure compensates for the lack of safeguards in the E-evidence proposals and provides additional oversight, partially solving the controversy of delegating decisions to service providers. As an additional safeguard, the proposal suggests that for cases where the order comes from a member state that is subject to proceedings initiated under the EU Treaty for serious and persistent breach of the EU values, the executing state should confirm the evidence release.

[44] De Busser, The Digital Unfitness, 174-175; Carrera, Stefan and Mitsilegas, I; EuroISPA, "Proposal for a Regulation on European Production and Preservation Orders for Electronic Evidence in Criminal Matters", (June 2018), https://rm.coe.int/ws1-1806-euroispa-e-evidence/16808c29c0 ; Meijers Committee, "Comments on the proposal for a regulation on European Production and Preservation Orders for electronic evidence in criminal matters", July 18, 2018, https://www.commissie-meijers.nl/comment/comments-on-the-proposal-for-a-regulation-on-european-production-and-preservation-orders-for-electronic-evidence-in-criminal-matters/

[45] EDRi, "EU 'e-evidence' proposals turn service providers into judicial authorities", April 17, 2018,
https://edri.org/our-work/eu-e-evidence-proposals-turn-service-providers-into-judicial-authorities /

[46] Valsamis Mitsilegas, "The Privatisation of Mutual Trust in Europe's Area of Criminal Justice: The Case of e-Evidence", *Maastricht Journal of European and Comparative Law 25, no. 3 (June 2018)*, 265

[47] European Parliament, "Report on the proposal for a directive of the European Parliament and of the Council laying down harmonised rules on the appointment of legal representatives for the purpose of gathering evidence in criminal proceedings", December 11, 2020, https://www.europarl.europa.eu/doceo/document/A-9-2020-0256_EN.html

The suggestions on notification procedures by the LIBE Committee can undoubtedly lead to raising the level of protection by involving more than one state in executing direct requests to providers. More importantly, this proposal represents a middle way between the current system of mutual legal assistance, which cannot meet the challenges of digital investigation, and direct requests mechanisms, which might be less cumbersome but entail too many risks for fundamental rights. The notification mechanism, if appropriately implemented, can reduce the dangers of privatisation of law enforcement and provide enough involvement of competent authorities of the affected state and executing state to guarantee that the direct requests will not be misused and abused.

5 Second Additional Protocol to the Budapest Convention

The Second Additional Protocol to the Budapest Convention was adopted by the Committee of Ministers of the Council of Europe on 17 November 2021 and opened for signature in May 2022.[48] Compared to the EU E-evidence proposal, the Protocol took a more moderate approach to enabling direct requests to providers located abroad: such requests can be sent only for domain name registration information (Article 6) and subscriber information (Article 7). The requests for domain registration data are not compulsory, and the provider can reject them. Therefore, Article 6 is trying to enhance the current mechanisms for voluntary cooperation. For subscriber information requests, a state can reserve the right not to apply this provision at all or require simultaneous notification. A reservation about subscriber information can also be made about specific types of data when it is "inconsistent with the fundamental principles of its domestic legal system". Other provisions of the Second Additional Protocol are trying to streamline more traditional cooperation mechanisms. Article 8 aims to expedite existing mutual legal assistance mechanisms for subscriber and traffic data requests by providing shorter deadlines and expedited procedures for handling requests. Lastly, Article 9 enables expedited disclosure of data in case of emergency.

[48] Council of Europe, Cybercrime: Council of Europe strengthens its legal arsenal Ref. DC 207(2021), November, 17, 2021, https://search.coe.int/directorate_of_communications/Pages/result_details.aspx?ObjectId=0900001680a48ca6

In the process of negotiations, the Second Additional Protocol provisions were criticised for enabling mechanisms for direct requests to providers about subscriber information. In their statement to one of the earlier drafts, civil society groups called for the involvement of Parties on both sides in the cases of subscriber information requests.[49] These concerns seem to be partially addressed in the final version of the Protocol, allowing states to make reservations or demanding simultaneous notification. There is still a question as to how states will implement these provisions and how the decisions of individual states to implement safeguards such as notification or not to apply the direct requests mechanisms at all can impact the future of cooperation.

Nevertheless, compared to the EU E-evidence proposals, the Second Protocol took a more balanced approach. This approach is necessary, considering the diversity in the national legislation of the signatories of the treaty and possible differences in human rights protection in various jurisdictions. Of course, nobody could expect that the Second Additional Protocol would go as far with the direct requests mechanisms as the EU E-evidence proposals or the US CLOUD Act due to its global nature, the diversity of the countries who are the parties of the Budapest Convention. The Second Additional Protocol has a great potential to streamline and improve the current mechanisms for cooperation while preserving fundamental rights and safeguards, and it seems to a large extent to have followed a careful approach to fixing the problems of traditional mutual legal assistance mechanisms rather than disregarding them.

It remains to be seen how the adoption of the Second Additional Protocol will impact cooperation between the EU member states and non-EU countries. Undoubtedly, it will increase the opportunity to use the Budapest Convention and the Second Additional Protocol as a basis for cooperation in addition to various bilateral agreements and UN treaties. In this regard, cooperation between South Korea and the EU countries can be improved if South Korea joins the Budapest Convention[50] and signs the Second Additional Protocol.

[49] EFF, "Joint Civil Society Response to Discussion Guide on a 2nd Additional Protocol to the Budapest Convention on Cybercrime", June 28, 2018, https://www.eff.org/document/joint-civil-society-response-discussion-guide-2nd-additional-protocol-budapest-convention

[50] Even though South Korea is not a party to the Budapest Convention, efforts have been made to join the treaty. For more discussion on this topic, please see the chapter of Gibum **Kim**.

6 Conclusion

Law enforcement agencies are increasingly facing challenges to access electronic evidence stored abroad, while traditional mutual legal assistance mechanisms for formal cooperation in criminal investigations are not meeting the growing demand to get data quickly and reliably. This ever-increasing demand also creates a growing pressure on service providers to voluntarily cooperate in digital investigations and even makes providers subject to mandatory orders from foreign authorities to hand over data without mutual legal assistance requests. The differences in national legislation concerning fundamental rights, safeguards, and the applicability of specific investigatory measures to particular types of crimes are exacerbated by legal uncertainly created by the patchwork of solutions related to mutual legal assistance and voluntary cooperation. This situation calls for solutions that would provide more clarity for everybody—law enforcement agencies, services providers, and, not less important, the users of services affected by the collection of data by law enforcement agencies.

The new solutions, such as the US CLOUD Act and the EU E-evidence proposals, while providing mechanisms for speeding up international cooperation, could create more problems in the long term by compromising fundamental rights and privatising law enforcement mechanisms. Undoubtedly, the mutual assistance legal mechanisms are slow, are cumbersome, and do not meet the everyday challenges of digital investigations. Still, they are in place to maintain mutual trust between the countries and protect fundamental rights by imposing control by competent authorities and judiciary in both states. Disregarding them completely might have an adverse effect on trust between users and providers, providers and states, and citizens' trust in criminal justice in the long run. It is crucial that the frameworks enabling direct requests from law enforcement to providers abroad still maintain the high level of scrutiny, be it a notification mechanism or other instruments that guarantee that decisions related to fundamental rights are not left in the hands of private companies. The service providers cannot, and shall not, fully replace competent authorities in criminal investigations.

Lastly, while the EU E-evidence proposals aim to enhance access to evidence in the EU, they do not fix the mechanisms for cooperation with the countries outside its borders. To improve the cooperation with third countries, the EU still has to explore bilateral and multilateral avenues, where the issues of trust and fundamental rights protection would be of utmost importance. Many criminal investigations rely on mutual legal assistance mechanisms, which must be improved and streamlined. Aligning the efforts to improve current mechanisms in the EU with global efforts such as the Second Additional Protocol to Budapest Convention while preserving a high level of safeguards is the key to future solutions that will improve access to e-evidence without compromising on fundamental rights.

References

Bąkowski, Piotr and Voronova, Sofija. "Electronic evidence in criminal matters. Briefing. EU Legislation in Progress". (March 2021). https://www.europarl.europa.eu/RegData/etudes/BRIE/2021/690522/EPRS_BRI(2021)690522_EN.pdf

Carrera, Sergio, Stefan, Marco and Mitsilegas, Valsamis. "Cross-border data access in criminal proceedings and the future of digital justice". (2020). https://www.ceps.eu/wp-content/uploads/2020/10/TFR-Cross-Border-Data-Access.pdf

Christakis, Theodore and Terpan, Fabien. "EU-US Negotiations on Law Enforcement Access to Data: Divergences, Challenges and EU Law Procedures and Options". International Data Privacy Law, 11 (2), (2021): 81-106.

Council of Europe, Cybercrime: Council of Europe strengthens its legal arsenal Ref. DC (2021), November, 17, 2021, https://search.coe.int/directorate_of_communications/Pages/result_details.aspx?ObjectId=0900001680a48ca6

Daskal, Jennifer. "A New UK-US Data Sharing Agreement: A Tremendous Opportunity, If Done Right". (2016). https://www.justsecurity.org/29203/british-searches-america-tremendous-opportunity/

Daskal, Jennifer. "Microsoft Ireland, the CLOUD Act, and International Lawmaking 2.0". *Stanford Law Review Online, Vol. 71. (*2018*)*

De Busser, Els. "The Digital Unfitness of Mutual Legal Assistance". Security and Human Rights 28, no. 1 (2017): 161-179

de Hert, Paul, Parlar, Cihan and Thumfart, Johannes. "Legal Arguments Used in Courts Regarding Territoriality and Cross-Border Production Orders: From Yahoo Belgium to Microsoft Ireland." *New Journal of European Criminal Law 9, no. 3 (September* 2018*): 326-352.*

EDPB. "Annex. Initial legal assessment of the impact of the US CLOUD Act on the EU legal framework for the protection of personal data and the negotiations of an EU-US Agreement on cross-border access to electronic evidence". (2019). https://edpb.europa.eu/sites/default/files/files/file2/edpb_edps_joint_response_us_cloudact_annex.pdf

EDPB. "Guidelines 2/2018 on derogations of Article 49 under Regulation 2016/679". (2018). https://edpb.europa.eu/ourwork-tools/our-documents/guidelines/guidelines-22018-derogations-article-49-under-regulation_en
EDRi. "EU 'e-evidence' proposals turn service providers into judicial authorities". April 17, 2018, https://edri.org/our-work/eu-e-evidence-proposals-turn-service-providers-into-judicial-authorities/
EFF. "Joint Civil Society Response to Discussion Guide on a 2nd Additional Protocol to the Budapest Convention on Cybercrime". June 28, 2018. https://www.eff.org/document/joint-civil-society-response-discussion-guide-2nd-additional-protocol-budapest-convention
EuroISPA. "Proposal for a Regulation on European Production and Preservation Orders for Electronic Evidence in Criminal Matters", (June 2018), https://rm.coe.int/ws1-1806-euroispa-e-evidence/16808c29c0
European Commission, "Commission staff working document. Impact assessment". April 17, 2018a. https://eur-lex.europa.eu/legal-content/EN/TXT/PDF/?uri=CELEX:52018SC0118&from=EN
European Commission. "Proposal for a Directive of the European Parliament and of the Council laying down harmonised rules on the appointment of legal representatives for the purpose of gathering evidence in criminal proceedings". April 17, 2018b. https://eur-lex.europa.eu/legal-content/EN/TXT/?uri=COM%3A2018%3A226%3AFIN
European Commission. "Proposal for a Regulation of the European Parliament and of the Council on European Production and Preservation Orders for electronic evidence in criminal matters". April 17, 2018c. https://eur-lex.europa.eu/legal-content/EN/TXT/?uri=COM%3A2018%3A225%3AFIN
European Commission. "Questionnaire on improving criminal justice in cyberspace Summary of Responses" (n.d.) https://ec.europa.eu/home-affairs/sites/default/files/e-library/documents/policies/organized-crime-and-human-trafficking/e-evidence/docs/summary_of_replies_to_e-evidence_questionnaire_en.pdf
European Commission. "Recommendation for a Council Decision authorising the opening of negotiations in view of an agreement between the European Union and the United States of America on cross-border access to electronic evidence for judicial cooperation in criminal matters". February 5, 2019a. https://eur-lex.europa.eu/legal-content/EN/TXT/?uri=CELEX%3A52019PC0070
European Commission. Criminal justice: Joint statement on the launch of EU-US negotiations to facilitate access to electronic evidence. September, 26, 2019b. https://ec.europa.eu/commission/presscorner/detail/en/STATEMENT_19_5890
European Parliament. "Report on the proposal for a directive of the European Parliament and of the Council laying down harmonised rules on the appoint-

ment of legal representatives for the purpose of gathering evidence in criminal proceedings". December 11, 2020. https://www.europarl.europa.eu/doceo/document/A-9-2020-0256_EN.html

European Union. "Agreement with the United States on mutual legal assistance". July 19, 2003. https://eur-lex.europa.eu/legal-content/EN/TXT/?uri=LEGISSUM%3Ajl0052

European Union. "Directive 2014/41/EU Regarding the European Investigation Order in criminal matters". (2014). https://eur-lex.europa.eu/legal-content/EN/TXT/?uri=celex%3A32014L0041

European Union. "EU in Brief. Values". (n.d.). https://europa.eu/european-union/about-eu/eu-in-brief_en

European Union. "Framework Agreement between the European Union and its Member States, on the one part, and the Republic of Korea, on the other part". January, 23, 2013. https://eur-lex.europa.eu/legal-content/EN/TXT/?uri=CELEX%3A22013A0123%2801%29

Hohmann, Mirko and Barnett, Sophie. "System Upgrade. Improving Cross-Border Access to Electronic Evidence". January 21, 2019. https://www.gppi.net/2019/01/21/system-upgrade

Kent, Gail. "The Mutual Legal Assistance Problem Explained". (2015). http://cyberlaw.stanford.edu/blog/2015/02/mutual-legal-assistance-problem-explained

Meijers Committee. "Comments on the proposal for a regulation on European Production and Preservation Orders for electronic evidence in criminal matters". July 18, 2018. https://www.commissie-meijers.nl/comment/comments-on-the-proposal-for-a-regulation-on-european-production-and-preservation-orders-for-electronic-evidence-in-criminal-matters/

Mignon, Emmanuelle. "The CLOUD Act: Unveiling European Powerlessness. September, 5, 2020. https://legrandcontinent.eu/fr/2020/09/05/the-cloud-act-unveiling-european-powerlessness/

Mitsilegas, Valsamis. "The Privatisation of Mutual Trust in Europe's Area of Criminal Justice: The Case of e-Evidence". *Maastricht Journal of European and Comparative Law 25, no. 3 (June* 2018*):* 263–65.

Rodriguez, Katitza. "The US CLOUD Act and the EU: A Privacy Protection Race to the Bottom". April 9, 2018. https://www.eff.org/fr/deeplinks/2018/04/us-cloud-act-and-eu-privacy-protection-race-bottom

Rojszczak, Marcin. "CLOUD act agreements from an EU perspective". *Computer Law & Security Review, Volume 38, (*2020*).*

SIRIUS Project. "EU Digital Evidence Situation Report. 2nd Annual Report". December 1, 2020, https://www.europol.europa.eu/publications-documents/sirius-eu-digital-evidence-situation-report-2nd-annual-report

Siry, Lawrence. "Cloudy Days Ahead: Cross-Border Evidence Collection and Its Impact on the Rights of EU Citizens." *New Journal of European Criminal Law 10, no. 3 (September 2019): 227–50.*

Stefan, Marco and González Fuster, Gloria. "Cross-border Access to Electronic Data through Judicial Cooperation in Criminal Matters State of the art and latest developments in the EU and the US". December 3, 2018. https://www.ceps.eu/ceps-publications/cross-border-access-electronic-data-through-judicial-cooperation-criminal-matters/

Tosza, Stanislaw. "Cross-Border Gathering of Electronic Evidence : Mutual Legal Assistance, Its Shortcomings and Remedies". In *Société numérique et droit pénal*, edited by Daniel Flore and Vanessa Franssen. (Bruylant: 2019): 269-285.

United States v. Microsoft Corp., 584 US ___ (2018). https://supreme.justia.com/cases/federal/us/584/17-2/

US Cloud Act. https://www.justice.gov/dag/page/file/1152896/download

Vazquez Maymir, Sergi. "Anchoring the need to revise cross-border access to e-evidence", *Internet Policy Review 9 (3), (*2020*).* https://doi.org/10.14763/2020.3.1495

Wahl, Thomas. "E-Evidence: Start of Negotiations on EU-US Agreement". January 12, 2020. https://eucrim.eu/news/e-evidence-start-negotiations-eu-us-agreement/

Woods, Andrew K. "Data Beyond Borders: Mutual Legal Assistance in the Internet Era". January 27, 2015. https://uknowledge.uky.edu/law_facpub/518/

PART III

International Order and Cross-Cutting Issues

Recent Trends in UN Cybersecurity Governance and South Korea-EU Cooperation

JoonKoo Yoo

1 Introduction

The international community's perception of cyber threats[1] has grown in the past ten years because of cyber incidents like the "Stuxnet" attack, the WannaCry ransomware attack, distributed denial of service (DDoS) attacks, the Snowden revelations, and the Sony hack. These incidents have led the United Nations (UN), the European Union (EU), the North Atlantic Treaty Organization (NATO), the Shanghai Cooperation Organization (SCO), G20, and other multilateral/regional organizations to discuss international security in cyberspace. The UN Group of Governmental Experts on Developments in the Field of Information and

[1] See Mason **Richey**, "Cyber offense dominance, regional dynamics, and middle power-led international cooperation" in this volume.

J. Yoo (✉)
Korea National Diplomatic Academy, Seoul, South Korea
e-mail: jkyoo88@mofa.go.kr

© The Author(s), under exclusive license to Springer Nature Switzerland AG 2022

G. Boulet et al. (eds.), *Cybersecurity Policy in the EU and South Korea from Consultation to Action*, New Security Challenges,
https://doi.org/10.1007/978-3-031-08384-6_12

Telecommunications in the Context of International Security (UNGGE) was established in 2004 to comprehensively discuss the issue of cyberspace governance and the applicability and interpretation of international law and norms. There are two groups with deeply diverging viewpoints—the group led by China and Russia and like-minded countries (LMCs) led by the US, EU, Japan, and South Korea. These groups disagree on issues such as the classification of international laws and norms, cyberterrorism and cybercrime, and data security relating to sovereignty and jurisdiction in cyberspace. In June 2017, the fifth session of the UNGGE could not reach a consensus report that would have brought additional clarity to how international law regulates cyberspace. This session stepped even back from earlier achievements. This failure casts doubt on the effectiveness of the entire UNGGE process. Against this backdrop, many states are calling for a thorough reform of the current GGE format, which could include changes in the participants, decision-making process, and agendas.

The UNGGE format served for twenty years to discuss the promotion and creation of norms of state behaviour in cyberspace. With Resolution (A/C.1/73/L.27/Rev.1) the Russian Federation proposed a change by convening in 2019 an Open-Ended Working Group (OEWG). Composed of multi-stakeholders and therefore including non-governmental organizations, companies, and research institutions this group should in turn develop the rules for responsible behaviour of states in cyberspace as well as modes of their implementation. Consensus based, this new OEWG should avoid the creation of "club agreements" and, unlike the UNGGE, encourage an inclusive, open, transparent, and democratic negotiation process, while fostering the norms-building capacity of every state. In accordance with its mandate the OEWG discussed existing and potential threats for information security and cooperative measures to address them; the further development of rules, norms, and principles of responsible behaviour of states; the applicability of international law to the use of ICTs by states; confidence-building measures; capacity building; and outreach in the form of a regular institutional dialogue with stakeholders under the auspices of the UN. South Korea and the EU could effectively support the process, in particular the key agenda items CBMs and norms-building.

Against such backdrop, this chapter unfolds in the following order. Section 2 examines the contemporary discussions on cybersecurity norms and laws. Section 3 discusses the progress and challenges in determining the format of future cybersecurity governance. In Section 4 provides an in-depth analysis of the previous cases of South Korea-EU cooperation at various levels in the context of cyber diplomacy. More specifically, Sect. 4 highlights the necessity and significance of their strategic partnership based on a shared interest in cyber resilience as well as international public and human rights law. Finally, Sect. 5 provides reflections on the current landscape of discussions on cybersecurity governance, the preferences and limits of South Korea's cyber diplomacy including the merits and challenges of its cooperation with the EU, as well as insights on how to expand this cooperation.

2 Discussions on Cybersecurity Norms and Laws[2]

The US, EU, and the LMCs hold the strong view that it is too early for a new treaty or an international organization as they do not want to compromise the current Internet model; they acknowledge the role of the UN. Russia and China hold the contrary view, arguing for a new international treaty governing cyberspace at the UN as well as eventually for a new international organization. The vast majority of the LMCs see a new treaty or organization as a long-term challenge at best. They prefer the practical implementation of the cyber norms contained in the GGE reports and making use of the existing UN-affiliated organizations and entities. Given the fifth UNGGE's failure to submit a conclusive report, discussions on changing the current GGE format and on how to increase its membership to the open- ended format will gain traction.[3]

The LMCs, while recognizing the usefulness of the UNGGE process, adopt the premise of strengthened non-legally binding and voluntary cyber norms, and state-centred decision-making on the basis of consensus; they prefer not to open up the group. This clashes with the intention of others wishing to set up a new international discussion forum.

In the discussion at the UN on the applicability of international law to cyberspace, the major issues are state sovereignty, human rights, the right

[2] See the chapter by Cuihong **Cai**.

[3] Remark by Dr Daniel Stauffacher, "UN GGE and UN OEWG: How to live with two concert UN Cybersecurity processes", Speech at the Jeju Peace Forum, 29 May 2019. https://ict4peace.org/activities/ict4peace-at-the-jeju-peace-forum-how-to-live-with-two-concurrent-un-cybersecurity-processes/.

of self-defence, international humanitarian law, state responsibility, and supply chain security.

The right of self-defence codified in Article 51 of the U.N. Charter, established in the wake of World War II, is challenged by the following factors: the lack of a clear-cut distinction between non-state actors and state actors; changing criteria for the exercise of self-defence; and the merging of traditional and non-traditional security issues.

Both processes—the OEWG and UNGGE—revealed sharp divisions between two blocs on the application of international law in cyberspace. A key sticking point is to what extent the norms recognized in the 2015 GGE Report will be reaffirmed. The LMCs claim that the UN Charter applies in its entirety to cyberspace, and existing international law including international humanitarian law, international human rights law, and the law of international responsibility therefore apply to cyberspace. The opposing camp lead by China and Russia is in favour of creating new legally binding international law governing cyberspace. China, Iran, Cuba, and others reject any language that could justify the use of force in cyberspace as well as the inclusion of international humanitarian law. As a compromise international humanitarian law, the right of self-defence, and state responsibility were moved from the main text to the annex.[4]

Disagreement also covers non-binding voluntary norms which should supplement existing international law given the unique features of cyberspace. The state-centred bloc, led by China and Russia, wanted to seize the opportunity to write a new set of binding global rules, including on data security and supply chain security, but this was rejected by the LMCs. The compromise clarified that norms do not replace or alter existing obligations or rights under binding international law, but rather offer additional specific guidance on what constitutes responsible state behaviour in the use of ICTs.[5]

[4] Dan Efrony, "The UN Cyber Groups, GGE and OEWG-A Consensus is Optimal, But Time is of the Essence", *Just Security*, July 16 2021. https://www.justsecurity.org/77480/the-un-cyber-groups-gge-and-oewg-a-consensus-is-optimal-but-time-is-of-the-essence/.

[5] Ibid.

3 Future Format of Cybersecurity Governance

The format of discussion remains controversial: while the state-centred group insists on the continuation of the OEWG under the auspices of the United Nations as the main consultative body on cybersecurity, the LMCs proposed a Programme of Action (PoA) for advancing responsible state behaviour in cyberspace. The LMCs refused linking the OEWG and the PoA. The opposing camp regards the PoA as an insufficiently discussed initiative by a few. China used this occasion to argue for the inclusion of its own initiative, Global Data Security, in the work programme of the future OEWG, at par with the PoA. The final OEWG report of 12 March 2021 notes that the PoA should be further elaborated at the new OEWG to be convened from 2021 to 2025.[6]

The failure of the 2016–2017 GGE to adopt a consensus report as well as the conflict between competing blocs has deepened scepticism about the effectiveness of cybersecurity discussions at the UN. Therefore, the adoption of the final OEWG report (2021) despite covid-19 is particularly significant, although the mentioned differences have not been overcome. The second OEWG (2021–2025), based on a Russian proposal, will have to navigate intense competition between the two blocs seeking to wrest control. The new OEWG mandate will determine the future course of the two parallel UN cyber discussion processes. If the OEWG becomes the key forum for cybersecurity deliberation that does substantive work, discussions on cybersecurity at the UN could be brought back on track. Decisions also need to be taken on whether to maintain the parallel structure with the PoA. As there are growing calls for an open-ended process that allows the participation of a broader range of countries, regardless of which bloc they belong to, the OEWG might pair up with the PoA in the years ahead, thereby replacing the existing dual-track discussions of the GGE/OEWG.

In order to be ready for the resumption of discussions in whatever format, South Korea needs to prepare. As an expert-driven structure is most likely, a pool of well-versed experts or a research consortium should be established in areas like policies, norms, and emerging technologies.

[6] JoonKoo Yoo, "UN Open-Ended Working Group Final Report: Issues and Implication", *IFANS Focus*, April 3, 2021. https://www.ifans.go.kr/knda/ifans/eng/pblct/PblctView.do?clCode=P11&pblctDtaSn=13784&koreanEngSe=ENG.

Moreover, as multi-stakeholders engagement is expected to be one of the defining features of the new process, Korean companies, research institutes as well as NGOs should be encouraged to participate.

4 South Korea-EU Cooperation in the Context of Cyber Diplomacy

4.1 South Korea's International Cooperation

South Korea has been actively involved in multilateral discussions on the applicability of international law in cyberspace, norms-building, confidence-building measures (CBMs), and capacity building. South Korea has participated in most of the six GGEs established since 1998 at the UN to discuss the threats posed by the use of ICTs in the context of international security. As a member of the 2015 GGE, South Korea endorsed the view that international law applies to cyberspace. The government's views on the applicability of international law have been primarily guided by the Seoul Framework for and Commitment to Open and Secure Cyberspace developed for the Global Conference on Cyberspace (GCCS) hosted in Seoul in 2013.[7] These views are widely shared by the EU and the LMCs.

In the previous GGE iterations, South Korea not only worked closely with the US but also coordinated its stance with others like Estonia, Finland, and Switzerland,[8] "regarded as champions of rule of law and potentially accepted as honest brokers in international cyber affairs".[9] For instance, South Korea, Estonia, Finland, and Switzerland insisted that due diligence should be included in the section on how international law applies in cyberspace—in contrast to the US which insisted on voluntary law. Driven by the imperative to address cybersecurity threats emanating

[7] The framework identified priorities for cyberspace governance (economic growth, social, and cultural benefits), international cybersecurity (promotion of voluntary CBMs and transparency measures), cybercrime (law enforcement cooperation in the investigation and prosecution of international cases), and capacity building (enhancement of efforts to close the digital divide). https://www.un.org/disarmament/wp-content/uploads/2019/10/ENCLOSED-Seoul-Framework-for-and-Commitment-to-an-Open-and-Secure--Cyberspace.pdf

[8] Statement by Republic of Korea, 2015 GGE, UN ODA, 2019.

[9] Eneken Tikk and Mika Kettunen, 2018, *Cyber Treaty is Coming*, p. 6. https://cpi.ee/wp-content/uploads/2018/02/Cyber-Treaty-is-Coming-Tikk-Kerttunen.pdf.

from North Korea, South Korea also used the GGE meetings to highlight the need to address the problem of cyber proxies.

South Korea also engaged in some of the private or public-private norm-building initiatives that proliferated following the failure of the GGE in 2017, such as the 2018 Paris Call for Trust and Security in Cyberspace at the UNESCO Internet Governance Forum (IGF) with other 63 states.[10] The purpose was to facilitate "upholding the multi-stakeholder approach in cybersecurity".

In 2019, South Korea had engaged in 13 bilateral cyber dialogues with major cyber powers such as Russia, the US, India, and Australia to bolster bilateral and trilateral cooperation.[11] Given South Korea's dependence on US security guarantees and the high sophistication of US cyber capacities, this dialogue carries weight. Both states established the South Korea-US Cyber Policy Consultations and the Cyber Cooperation Working Group between their militaries as well as a South Korea-US ICT Policy Forum. They also hold joint cyber simulation exercises. The US and South Korea are both committed to the multi-stakeholder Internet governance model and to increasing cyber cooperation by "(1) enhancing information sharing on cyber threats, particularly to critical infrastructure; (2) strengthening collaboration on investigation on cyber incidents; (3) deepening military-to-military cyber cooperation; and (4) encouraging collaboration on cybersecurity research and development, education and workforce development, and cooperation on technology between cybersecurity industries".[12] A Blue House-White House cyber coordination channel at the highest level underscores the importance.

With the elections of new leaders in the US and South Korea in 2016 and 2017, respectively, cybersecurity cooperation became less central. In a joint statement in 2017, both sides briefly stated that "advancing cooperation on cybersecurity is required for economic growth, enhancing the Japan-South Korea-US trilateral relationship to address cybersecurity

[10] France Diplomacy, "Cybersecurity: Paris Call of 12 November 2018 for Trust and Security in Cyberspace", Press Room Februaru 2021. https://www.diplomatie.gouv.fr/en/french-foreign-policy/digital-diplomacy/france-and-cyber-security/article/cybersecurity-paris-call-of-12-november-2018-for-trust-and-security-in.

[11] JoonKoo Yoo, Interview, the German Marshall Fund, March 21, 2019.

[12] The White House, 2015. Joint Fact Sheet. The United States-Republic of Korea Alliance: Shared Values, New Frontiers, October 16, 2015. https://obamawhitehouse.archives.gov/the-press-office/2015/10/16/joint-fact-sheet-united-states-republic-korea-alliance-shared-values-new.

issues, and to collectively respond to North Korea's dangerous and destabilizing malicious cyber activity".[13]

The Moon government's efforts to improve relations with China and North Korea necessitated maintaining greater distance to the US and Japan. Former US President Donald Trump's unilateral call to cancel joint military exercises added uncertainty and motivated the South Korean government to invest more in its own cyber resilience.

Finally, six years after a trilateral summit in 2008 had decided to intensify cybersecurity cooperation through an Action Plan for Promoting Trilateral Cooperation, Seoul co-launched the China-Japan-South Korea trilateral cybersecurity dialogue in 2014. These then annual dialogues at Director General's level served primarily as information sharing platforms. Complimentary to this Korean bilateral and trilateral cyber diplomacy, regional organizations in East Asia have increased regional cybersecurity cooperation as global multilateral negotiations stalled. Concerns about the growing cyber capabilities of North Korea contributed to this regional development. South Korea has been most active in discussions at the ASEAN Regional Forum (ARF), which has a working group on cybersecurity. Following the ARF Statement on Cooperation in Ensuring Cybersecurity,[14] adopted during the 19th ARF in 2012, the ARF published a work plan on cybersecurity cooperation in May 2015 and held a series of meetings in 2015 and 2016. Since 2015, the Council for Security Cooperation in the Asia Pacific (CSCAP), the ARF's non-governmental Track 2.0 dialogue mechanism based in Kuala Lumpur, also hosted several study groups on developing cyber norms of behaviour and CBMs for Asia Pacific to promote the development of regional cyber norms based on a multi-stakeholder model. The latest study group launched in mid-2018 and chaired by Singapore has sought to further develop previous recommendations to inaugurate an ARF ICT Security Inter-Sessional Meeting (ISM) and/or an ARF Study Group on CBMs. A report dealt with the

[13] The White House, 2017. Joint Statement between the United States and the Republic of Korea, June 30, 2017. https://www.whitehouse.gov/briefings-statements/joint-statement-united-states-republic-korea /. Cp. David Sungjae Hong, "Impact Player: Moon Jae-in", *CSIS*, April 25, 2017. https://www.csis.org/analysis/impact-player-moon-jae; Eleanor Albert, "South Korea's Chaebol Challenge", *Council on Foreign Relations*, 4 May 2018. https://www.cfr.org/backgrounder/south-koreas-chaebol-challenge.

[14] 19th ASEAN Regional Forum (ARF) Ministerial Meeting, Statement by the Ministers of Foreign Affairs on Cooperation in Ensuring Cyber Security, July 12, 2012. https://www.diplomacy.edu/wp-content/uploads/2021/11/Asean_Regional_Forum_Statement.pdf.

implementation and improved the management of regional cybersecurity CBMs.[15]

South Korea supported the CBMs agreed at the ARF and by the Organization for Security and Cooperation in Europe (OSCE) as the ideal models for advanced regional cooperation,[16] helping "to promote trust and assurance in cyberspace across the world".[17] In a 2015 national position paper, South Korea stated that in the absence of international frameworks on CBMs, "it is important to establish measures to build confidence in the meantime, which will increase predictability and reduce misperception, to help reduce the risk of conflict".[18] In 2017 and 2019, South Korea and the OSCE jointly organized Inter-Regional Conferences on Cyber/ICT Security in Seoul to share best practices in trust-building in ASEAN and Europe and to facilitate regional CBMs.[19]

4.2 *EU-South Korea Strategic Partnership and Shared Interests in Cyber Resilience*

As one of the EU's 11 strategic partners (five of which are in Asia), the EU recognized South Korea as a country with whom cooperation is instrumental "to contribute to the resolution of international crises and to address the key challenges of the 21st century".[20] In the area of foreign and security policy, the EU and South Korea have cooperated bilaterally as well as in interregional and global multilateral institutions on a wide range

[15] Council for Security Cooperation in the Asia Pacific (CSCAP), 2019. "Developing cyber norms of behaviour and confidence building measures for Asia Pacific", CSCAP Study Group on International Law and Cyberspace. http://www.cscap.org/index.php?page=developing-cyber-norms-of-behaviour-and-confidence-building-measures-for-asia-pacific.

[16] Ibid.

[17] South Korea, Ministry of Foreign Affairs Press Release "Inter-Regional Conference on Cyber/ICT Security to Take Place", 2017. https://www.mofa.go.kr/eng/brd/m_5676/view.do?seq=318297&srchFr=&srchTo=&srchWord=Conference&srchTp=0&multi_itm_seq=0&itm_seq_1=0&itm_seq_2=0&company_cd=&company_nm=&page=3&titleNm.

[18] Ibid., p. 5.

[19] Already in 2014, South Korea hosted the "Asia-Pacific Regional Seminar on International Law and State Behaviour in Cyberspace" together with the UN Institute for Disarmament Research (UNIDIR).

[20] Delegation of the European Union to the Republic of Korea, "The Republic of Korea and the EU", May 10, 2016. https://eeas.europa.eu/delegations/south-korea_en/8789/The%20Republic%20of%20Korea%20and%20the%20EU.

of issues. On 30 June 2020, President Moon Jae-in, European Council President Charles Michel, and European Commission President Ursula von der Leyen held a leaders' meeting via video conference. In their joint press statement, they underlined their commitment "to strengthen cooperation in responding to the Fourth Industrial Revolution, harnessing human-centric digitalization and promoting safe and ethical implementation of Artificial Intelligence technologies". Both sides agreed on the launch of the High-Level Dialogue on digital economy in the near future and confirmed their commitment to promote the rules-based international order and effective cooperation in the fields of cybersecurity, disinformation, and additional hybrid threats.[21] Such measures reflect the shared interest of the EU and South Korea in an international rules-based order in cyberspace in line with international public and human rights law. Both have invested significantly in emerging technologies in areas such as AI and 5G and are therefore interested in setting standards and sharing best practices in building competitive digital economies. The EU's normative power status has recently gained greater attention through the General Data Protection Regulation (GDPR): on 17 December 2021, the European Commission adopted the adequacy decision ensuring safe data flows between the EU and South Korea.[22] To this end, South Korea had adapted its legislation to meet GDPR requirements.[23]

Third, both partners benefit from exchanging best practices on cyber defence and resilience to meet cyber threats.[24] While threats from Russian

[21] European Commission, "Joint press release: Republic of Korea – EU Leaders' video conference meeting", June 30, 2020, available at https://ec.europa.eu/commission/presscorner/detail/en/IP_20_1226.

[22] See the announcement of the European Commission: https://ec.europa.eu/commission/presscorner/detail/en/statement_21_6915; see also the chapter by George **Christou** and Ji Soo **Lee**.

[23] In December 2018, South Korea's Act on the Promotion of IT Network Use and Information Protection (Network Act) was updated, requiring digital communications providers handling South Korean data to establish a domestic representative in South Korea overseeing data protection issues from March 2019 onwards. In January 2020, the Korean parliament passed another set of amendments to the Network Act, the Personal Information Protection Act (PIPA), and the Act on the Use and Protection of Credit Information (Credit Information Act). See "Korea introduces major amendments to data privacy laws", *Lexology*, January30,2020,availableathttps://www.lexology.com/library/detail.aspx?g=223051e2-0346-4935-8fbb-6146e4424d94.

[24] Mason Richey, "EU-South Korea Security Relations: The current state of play", *Egmont Security Policy Briefs*, May 30, 2017, p. 4. https://www.egmontinstitute.be/eu-south-korea-security-relations-the-current-state-of-play/.

territory constitute the most immediate challenge for Europe's cybersecurity, European institutions and banks have also been victims of attacks originating from North Korea such as the WannaCry hack. Potentially more significant in the long run are threats emanating from China, either through companies or through the Chinese military's technological build-up. These concerns range from economic and national security espionage to the coercive potential of China's comparative military advantage.

Fourth, as mentioned above, both sides share an interest in improving cooperation on countering cybercrime. Passing the necessary legal changes and finishing the ratification of the Budapest Convention will be an important contribution by South Korea.

Since the 2015 Summit South Korea and the EU "agreed to increase bilateral cooperation on cyberspace as well as to strengthen the global partnership in response to threats arising from cyberspace".[25] These shared political and economic interests related to cyber resilience and diplomacy have further deepened cyber policy cooperation between the EU and South Korea.[26]

4.3 South Korea's Cyber Policy Cooperation with the EU

Building on these convergent interests, the EU and South Korea institutionalized a political dialogue on cybersecurity. The 2010 Framework Agreement had already included a paragraph on combating cybercrime.[27] At the 2015 EU-South Korea Summit, leaders "emphasized the importance of ensuring the openness and security of cyberspace for it to continue being a driving force for the freedom, prosperity and economic growth of mankind" and "agreed to increase bilateral cooperation on cyberspace as well as to strengthen the global partnership in response to threats arising from cyberspace", including by enhanced cooperation

[25] The EU and South Korea, "Joint Press Statement, 8th Republic of Korea-EU Summit", The European Council, September 15, 2015, p. 3. http://www.consilium.europa.eu/media/24474/korea-eu-summit-15_09_2015.pdf.

[26] For further details on cyber policy cooperation between the EU and South Korea, see the chapter by George **Christou** and Ji Soo **Lee**.

[27] Thomas Renard, "EU cyber partnerships: assessing the EU strategic partnerships with third countries in the cyber domain," European Politics and Society, January 28, 2018, p. 15. http://www.egmontinstitute.be/content/uploads/2018/01/EPS-EU-cyber-partners_RENARD_AM.pdf?type=pdf.

between the Kr-CERT and EU-CERT.[28] Between 2015 and early 2020, five Cyber Dialogue meetings were held to exchange information on cyber threat perceptions, respective cyber policies, and views on international issues such as the applicability of international law, norms of responsible state behaviour and CBMs.[29] The dialogues also covered issues related to the promotion and protection of human rights online, including freedom of expression, cyber capacity building efforts in third countries, and research and development on cyber issues. Despite a high degree of convergence, the potential to act instead of simply exchanging information and building trust[30] was not realized as there was no sense of urgency.[31] Nevertheless, the evolving EU cybersecurity policymaking and establishing greater cooperation at the operational level, for example, between the NIS and ENISA or NATO, raised interest within the South Korean intelligence and strategic community.

In 2018, the European Commission funded the three-year EU Cyber Direct project to expand the EU's cyber dialogue with strategic partners, including South Korea, by linking governmental and non-governmental expertise on norms, CBMs, and capacity building.[32] In a policy paper on enhanced EU security cooperation in and with Asia, the EU also noted that it aims to "enhance cooperation [with Asia] in the field of cybersecurity in favour of a global, open, free, stable and secure cyberspace".[33]

Finally, in the field of digital economy and emerging technologies, cooperation between the EU and South Korea is centred around future communication networks, 5G, cloud, IoT, Internet trust and security, policies to support start-ups, and e-governance.[34]

[28] European Council, Joint Statement, 8th Republic of Korea-EU Summit, September 15, 2015. https://www.consilium.europa.eu/media/24474/korea-eu-summit-15_09_2015.pdf.

[29] See "4th European Union – Republic of Korea Cyber Dialogue held in Seoul", European External Action Service, January 30, 2018. https://eeas.europa.eu/delegations/south-korea/38995/4th-european-union-republic-korea-cyber-dialogue-held-seoul_en.

[30] EU official, cited in Thomas Renard, "EU Cyber Partnerships", p. 15.

[31] Interview, EU official, Seoul, March 19, 2019.

[32] The author was affiliated to the EU Cyber Direct project at the time of writing.

[33] "Remarks by EU Ambassador Michael Reiterer for 2018 Seoul International Forum on Elections", European External Action Service, June 11, 2018. https://eeas.europa.eu/delegations/south-korea/46160/remarks-eu-ambassador-michael-reiterer-2018-seoul-international-forum-elections_en.

[34] See also the chapter by George **Christou** and Ji Soo **Lee**.

5 Conclusion

Advances in science and technology often drive the growth of a nation's cyber capabilities, but there are also dark sides to such advances. Cyber threats are becoming increasingly sophisticated and diversified, requiring countries around the world to build a common understanding of new types of cyber threats. Despite the broad consensus on the causes of cyber threats, countries have dissonant views on the guiding principles of cybersecurity policies. The like-minded countries (LMCs) stressed that their concerns are not about cyber capabilities themselves, but about their malicious use, which poses threats to international peace and security. The LMCs, therefore, suggested that the GGE/OEWG report should stipulate the notion of "technological neutrality". China, on the other hand, held that the notion of "technological neutrality" is unacceptable, although it supports the idea of advancing the notion of "technology for good".

Technology-prone states are faced with the so-called capability-vulnerability paradox—the more its economy, governance, and society are digitized, the more vulnerable it becomes.

Three impediments weaken the state's ability to effectively manage this paradox. First, as in other democratic and quickly digitizing states, measures to increase resilience have often been undermined by bureaucratic politics and competition among the various institutions involved and a concomitant reluctance to coordinate and share information. Second, the difficulty to combat cyberattacks exposes the limitations of deterrence in cyberspace, which is less effective com- pared to conventional domains since attribution and signalling become more perplex because of the use of "false flags" and proxies. Finally, the success of South Korea's cyber resilience efforts has been constrained by deteriorating security developments in Northeast Asia. During the past decade, conventional interstate rivalries recurrently escalated, and the mounting use of ICTs in these rivalries increased the risks of misperception, miscommunication, miscalculation, and, as a result, conflict escalation in the region, in which CBMs remain underdeveloped.

This chapter also outlined the preferences and limits of South Korea's multilateral, regional, bilateral, and plurilateral cyber diplomacy, and the promises and perils of its cooperation with the EU on ICT and research, Internet governance, cybercrime, and cyber norms. Domestic national security concerns have strongly driven South Korea's active cyber diplomacy, which, like its broader foreign policy orientation, primarily focuses on regional developments. This applies in particular to North Korea's cyber capabilities and often trumped concerns related to data protection and privacy. Still, South Korea's views on the applicability of international law and the need to swiftly implement agreed-upon norms, advance CBMs, and jointly build cybersecurity capacity largely converge with those of the EU. In addition, both sides share broader political and economic interests related to cyber diplomacy and resilience such as the dependence on a rules-based order and standards that define and deter irresponsible behaviour in cyberspace and guarantee normative transparency and supply chain security in the context of their investments in emerging technologies.

These convergences were facilitated by interactions in the Cyber Dialogue and ICT Dialogue between the EU and South Korea in the past and can incentivize broader cybersecurity cooperation in the future. Both sides can use the high-level political recognition of the need of global cooperation on cyber resilience in the context of the covid-19 pandemic to double down on their efforts to address cyber conflict and warfare, disinformation, and crime. In the field of cyber conflict prevention, collaborating towards a successful outcome of the OEWG that consolidates previous gains and outlines a future framework for dialogue on the use of ICTs in the context of international security would be an immediate priority.

Subsequently, both sides can build on the interregional exchanges between the ARF and the OSCE to further their collaboration on CBMs and probe extended cooperation of the EU with the South Korea's Global Cybersecurity Center for Development (GCCD)[35] or The Cybersecurity Alliance for Mutual Progress (CAMP)[36] in the area of cybersecurity capacity building. In the field of disinformation, both sides can exchange best practices on the regulation of disinformation during the pandemic, which underscored the potentially lethal effects of widespread false information. In the field of cybercrime, both sides can coordinate their diplomatic efforts in the ongoing negotiations on a new cybercrime treaty in the UNGA's Third Committee. Yet, the pandemic made clear that a broadened dialogue on cybersecurity cannot be decoupled from discussions on data protection and privacy. South Korea's use of surveillance technologies to mitigate the pandemic might be issues during the periodic review

[35] Korea Internet & Security Agency (KISA), Global Project (GCCD). https://www.kisa.or.kr/EN/201.

[36] Korea Internet & Security Agency (KISA), Cybersecurity Alliance for Mutual Progress (CAMP). https://www.cybersecalliance.org/camp/about.do.

of the adequacy decision. Building a secure cyberspace that is free and open will require the persistent engagement of multiple stakeholders in the diverse cyber policy landscape in Europe and South Korea.

References

"Remarks by EU Ambassador Michael Reiterer for 2018 Seoul International Forum on Elections". European External Action Service, June 11, 2018. https://eeas.europa.eu/delegations/south-korea/46160/remarks-eu-ambassador-michael-reiterer-2018-seoul-international-forum-elections_en.

Council for Security Cooperation in the Asia Pacific (CSCAP). 2019. "Developing cyber norms of behaviour and confidence building measures for Asia Pacific", CSCAP Study Group on International Law and Cyberspace. http://www.cscap.org/index.php?page=developing-cyber-norms-of-behaviour-and-confidence-building-measures-for-asia-pacific.

Daniel Stauffacher. "UN GGE and UN OEWG: How to live with two concert UN Cybersecurity processes", Speech at the Jeju Peace Forum, May 29, 2019. https://ict4peace.org/activities/ict4peace-at-the-jeju-peace-forum-how-to-live-with-two-concurrent-un-cybersecurity-processes/.

Dan Efrony. "The UN Cyber Groups, GGE and OEWG-A Consensus is Optimal, But Time is of the Essence", *Just Security*, July 16, 2021. https://www.justsecurity.org/77480/the-un-cyber-groups-gge-and-oewg-a-consensus-is-optimal-but-time-is-of-the-essence/.

David Sungjae Hong. "Impact Player: Moon Jae-in", *CSIS*, April 25, 2017. https://www.csis.org/analysis/impact-player-moon-jae.

Delegation of the European Union to the Republic of Korea. "The Republic of Korea and the EU", May 10, 2016. https://eeas.europa.eu/delegations/south-korea_en/8789/The%20Republic%20of%20Korea%20and%20the%20EU.

Eleanor Albert. "South Korea's Chaebol Challenge", *Council on Foreign Relations*, 4 May 2018. https://www.cfr.org/backgrounder/south-koreas-chaebol-challenge.

Eneken Tikk and Mika Kettunen. 2018. *Cyber Treaty is Coming*. https://cpi.ee/wp-content/uploads/2018/02/Cyber-Treaty-is-Coming-Tikk-Kerttunen.pdf.

European Commission. "Landmark agreement between the European Commission and South Korea on 5G mobile technology", June 16, 2014. http://europa.eu/rapid/press-release_IP-14-680_en.htm.

European Commission. "Joint press release: Republic of Korea – EU Leaders' video conference meeting", June 30, 2020a. https://ec.europa.eu/commission/presscorner/detail/en/IP_20_1226.

European Commission. *Shaping Europe's digital future. Asia-Pacific*, 2020b. https://ec.europa.eu/digital-single-market/en/asia.

European Council. Joint Statement, 8th Republic of Korea-EU Summit, September 15, 2015. https://www.consilium.europa.eu/media/24474/korea-eu-summit-15_09_2015.pdf.

France Diplomacy. "Cybersecurity: Paris Call of 12 November 2018 for Trust and Security in Cyberspace", Press Room Februaru 2021. https://www.diplomatie.gouv.fr/en/french-foreign-policy/digital-diplomacy/france-and-cyber-security/article/cybersecurity-paris-call-of-12-november-2018-for-trust-and-security-in.

JoonKoo Yoo. "UN Open-Ended Working Group Final Report: Issues and Implication", *IFANS Focus*, April 3, 2021. "UN Open-Ended Working Group Final Report: Issues and Implication". https://www.ifans.go.kr/knda/ifans/eng/pblct/PblctView.do?clCode=P11&pblctDtaSn=13784&koreanEngSe=ENG.

"Korea introduces major amendments to data privacy laws". *Lexology*, January 30, 2020. https://www.lexology.com/library/detail.aspx?g=223051e2-0346-4935-8fbb-6146e4424d94.

Statement by Republic of Korea. 2015 GGE, UN ODA, 2019.

Mason Richey. "EU-South Korea Security Relations: The current state of play", *Egmont Security Policy Briefs*, May 30, 2017, p. 4. https://www.egmontinstitute.be/eu-south-korea-security-relations-the-current-state-of-play/.

South Korea. Ministry of Foreign Affairs Press Release "Inter-Regional Conference on Cyber/ICT Security to Take Place", 2017. https://www.mofa.go.kr/eng/brd/m_5676/view.do?seq=318297&srchFr=&srchTo=&srchWord=Conference&srchTp=0&multi_itm_seq=0&itm_seq_1=0&itm_seq_2=0&company_cd=&company_nm=&page=3&titleNm.

"4th European Union – Republic of Korea Cyber Dialogue held in Seoul". European External Action Service, January 30, 2018. https://eeas.europa.eu/delegations/south-korea/38995/4th-european-union-republic-korea-cyber-dialogue-held-seoul_en.

The EU and South Korea. "Joint Press Statement 8th Republic of Korea-EU Summit", The European Council, September 15, 2015. http://www.consilium.europa.eu/media/24474/korea-eu-summit-15_09_2015.pdf.

The White House. 2015. Joint Fact Sheet. The United States-Republic of Korea Alliance: Shared Values, New Frontiers, October 16, 2015. https://obamawhitehouse.archives.gov/the-press-office/2015/10/16/joint-fact-sheet-united-states-republic-korea-alliance-shared-values-new.

The White House. 2017. Joint Statement between the United States and the Republic of Korea, June 30, 2017. https://www.whitehouse.gov/briefings-statements/joint-statement-united-states-republic-korea.

Thomas Renard. "EU cyber partnerships: assessing the EU strategic partnerships with third countries in the cyber domain," European Politics and Society, January 28, 2018. http://www.egmontinstitute.be/content/uploads/2018/01/EPS-EU-cyber-partners_RENARD_AM.pdf?type=pdf.

The International Cybersecurity Cooperation Dilemma and Implications for EU-South Korea Relations

Cuihong Cai

1 Introduction

Cyberspace eliminates physical distance between countries, increases the frequency of interactions between actors, and makes the world more connected. Nevertheless, it is often a trouble spot in international relations. It can increase the frequency of conflicts between countries, reduce the degree of mutual trust, and exacerbate national insecurity. Cybersecurity cooperation is affected by bilateral and multilateral relations. As a result, it faces challenges in terms of the cooperation dilemma. Studying the dilemma of cybersecurity cooperation is important for promoting bilateral security cooperation, exploring a benign cooperation model among major powers, and establishing an international cybersecurity cooperation mechanism.

C. Cai (✉)
Fudan University, Shanghai, China
e-mail: chcai@fudan.edu.cn

© The Author(s), under exclusive license to Springer Nature Switzerland AG 2022

G. Boulet et al. (eds.), *Cybersecurity Policy in the EU and South Korea from Consultation to Action*, New Security Challenges,
https://doi.org/10.1007/978-3-031-08384-6_13

This chapter consists of five parts. The first summarizes models of international cybersecurity cooperation based on the subject of governance and the scope of participation. The second part discusses problems of international cybersecurity cooperation including fragmented mechanisms, poor effectiveness, and camp-based confrontation. The third part analyzes the reasons for difficulties in cybersecurity cooperation, both subjectively and objectively. The fourth part discusses key issues in EU-South Korea cybersecurity cooperation and the last part examines the policy direction of EU-South Korea cybersecurity cooperation. This chapter advocates inclusive and open cooperation to increase cybersecurity. The EU and South Korea could consider developing a tailor-made framework for cybersecurity cooperation and consolidating existing but parallel cybersecurity cooperation into a dedicated cybersecurity cooperation mechanism. In addition, Europe and South Korea need to carefully consider the possible impact of the confrontation between camps, and it is necessary to pursue greater autonomy in cyberspace.

2 Models of International Cybersecurity Cooperation

International cooperation in cybersecurity involves government and non-governmental organizations. At present, the non-governmental governance structure has few disputes and is focused on infrastructure and technical aspects of the Internet. Therefore, the focus of this chapter is on intergovernmental cybersecurity cooperation. This can be divided into three models according to the scope of participation: (1) a global cooperation model under the framework of international organizations such as the United Nations; (2) a regional cooperation model under the framework of regional organizations represented by the European Union, NATO, the Organization for Economic Cooperation and Development (OECD), Asia Pacific Economic Cooperation (APEC) and the Shanghai Cooperation Organization (SCO); (3) a bilateral cooperation model including cybersecurity dialogue mechanisms and industry exchange platforms. As cyberspace is borderless, cybersecurity cooperation is a global issue. Therefore, effective bilateral and regional cybersecurity cooperation should aid global cooperation mechanisms.

Another categorization of international cybersecurity cooperation mechanisms is according to the tools used: international conference mechanisms, international organization mechanisms, international treaties, and soft law. International conferences and international organizations generally provoke discussions and reflections, eventually resulting in international treaties or international law.

Among international conference mechanisms, the most influential ones include the World Summit on the Information Society (WSIS) under the UN framework, the Internet Governance Forum (IGF), the United Nations Governmental Group of Experts (UNGGE), the UN Intergovernmental Expert Group on Cybercrime (IEG), and the Global Cybersecurity Agenda (GCA). In addition, communication platforms initiated by cybersecurity stakeholders or scientific research institutions also receive attention. Examples include the London Process, the Internet Society (ISOC), the World Internet Conference (WIC), the Commonwealth Internet Governance Forum (CIGF), and the International Multilateral Partnership against Cyber Terrorism (IMPACT). Moreover, the US-led Global Counter-Terrorism Forum (GCTF), the International Cybersecurity Protection Alliance (ICSPA) supported by the UK, the International Information Security Expert Group established by the SCO, and the Working Group on International Law in Cyberspace of the Asian-African Legal Consultative Organization (AALCO) are also regularly held seminars.[1]

In terms of international organization mechanisms, organizations such as the UN, the EU, and the Organization for Security and Cooperation in Europe (OSCE) deal with cybersecurity. International cooperation mechanisms under the UN framework such as the International Telecommunication Union (ITU), the WSIS, the United Nations Commission on Science and Technology for Development (CSTD), and the International Criminal Police Organization (ICPO) play an important role as well.[2] In addition, the European Cybercrime Center, initiated by the European Commission, was formally established in 2013. The OECD

[1] Among these, WSIS, IGF, UNGGE, GCCS, London Process, ISOC, WIC and CIGA are comprehensive meetings to discuss international Internet governance, cybersecurity, and international law in cyberspace; IEG, GCA, ICSPA, SCO, and AALCO merely focus on cybercrime; CIGA, GCTF, and IMPACT focus on cyberterrorism.

[2] The ITU is a specialized agency of the United Nations responsible for all matters related to information and communication technologies; CSTD, a subsidiary body of the Economic and Social Council (ECOSOC), examines specific issues on science and technology for development; ICPO is an international organization that facilitates worldwide police cooperation and crime control.

put forward the "Guidelines for the Security of Information Systems and Networks: Towards a Culture of Security"[3]and established an Action against Terrorism Unit. NATO established a Center of Excellence on Cyber Defense in 2008 and a Cyber Rapid Reaction Team in 2013. These international organizations provide long-term mechanisms for solving cybersecurity issues, acting as powerful swords against cybersecurity threats.

International treaties aim to bring about legally binding documents, which is an ideal solution to cybersecurity issues. In cyberspace, there is still a lack of international laws to restrict and punish vicious behaviors. Regulations are still at an early stage. At present, an important international treaty in cybersecurity is the Convention on Cybercrime under the auspices of the Council of Europe. The international community is also developing soft laws. These are not legally binding but are usually the result of tedious negotiations and accepted by sovereign states. One attempt within academia and by non-state actors to draft an international agreement is the Stanford Proposal for an International Convention on Cyber Crime and Terrorism.[4] Another example is the International Code of Conduct for Information Security, proposed by China, Russia, and several others, in contrast to the Convention on Cybercrime that reflects the interests of developed countries. Nevertheless, there remains a lack of globally recognized international laws in cyberspace.

3 Problems in International Cybersecurity Cooperation

With the development of the Internet, countries have gradually formed a set of methods for international cybersecurity cooperation. Despite this, the progress of international cybersecurity cooperation is still unsatisfactory. Today's international cybersecurity cooperation presents problems of mechanism fragmentation, poor effectiveness, and camp-based confrontation.

[3] OECD, *OECD Guidelines for the Security of Information Systems and Networks: Towards a Culture of Security*, OECD.org, accessed June 25, 2021, https://www.oecd.org/digital/ieconomy/oecdguidelinesforthesecurityofinformationsystemsandnetworkstowardsacultureofsecurity.htm.

[4] Abraham D. Sofaer, Seymour E. Goodman, *Proposal for an International Convention on Cyber Crime and Terrorism*, CISAC, accessed June 25, 2021, https://cisac.fsi.stanford.edu/publications/proposal_for_an_international_convention_on_cyber_crime_and_terrorism_a.

3.1 Mechanism Fragmentation

Diversified and multi-level international cybersecurity cooperation platforms provide opportunities for exchange but also bring about the problem of mechanism fragmentation, resulting in a variety and plurality of mechanisms, the blurring of boundaries of topic areas, the diversification of actors, and confusion related to governance roles and responsibilities.[5] Cybersecurity issues are usually discussed in multilateral mechanisms, most of which have diversified topics of interest and incorporate cybersecurity issues into existing discussions as opposed to professional cyberspace organizations that merely handle cybersecurity issues. For example, the APEC, the ASEAN Regional Forum (ARF), and the Asia-Pacific Telecommunity (APT) are all engaged in cybersecurity policy. Mechanism fragmentation may lead to increasing ambiguity, interference, and consequently ineffectiveness of cybersecurity rules. The EU's external cybersecurity cooperation mechanism is an example. The EU cooperates with some countries under the framework of the Common Security and Defense Policy (CSDP), with some regarding cyberterrorism and with another group of countries on cybercrime. The overall coherence of these policies, as well as their visibility is yet to be ensured.[6]

At this stage, there is no clear function allocation among these international mechanisms nor is there an effective means for coordination. Groups with different interests will easily switch from one platform to another when their needs are not met (forum shopping). For example, as the Convention on Cybercrime ignored the interests of developing countries, some developing countries represented by China and Russia turned to the United Nations and put forward the International Code of Conduct for Information Security under the framework of the United Nations accordingly.

The operating mechanisms and the compulsory force of international laws at different platforms also vary. For instance, although they are all international conventions on cybercrime, the Council of Europe

[5] Eileen Donahoe, "Governance Challenges in the Global Digital Ecosystem," In *Innovations in Global Governance: Peace-Building, Human Rights, Internet Governance and Cybersecurity, and Climate Change, Council on Foreign Relations*, accessed April 5, 2021, 24–28, https://www.cfr.org/report/innovations-global-governance.

[6] Thierry Tardy, "Revisiting the EU's Security Partnerships," European Union Institute for Security Studies (EUISS), accessed April 5, 2021, https://www.iss.europa.eu/content/revisiting-eu%E2%80%99s-security-partnerships.

Convention on Cybercrime, the Arab Convention on Combating Information Technology Offences, and the draft African Union Convention on Cybercrime have different scope of contracting parties and different enforcement powers. Moreover, most of the existing cybersecurity cooperation discussions are interspersed with other topics. This lack of focus renders dialogue and cooperation difficult.

3.2 Poor Effectiveness

The effectiveness of international mechanisms is analyzed on the basis of three criteria: (1) substantive content/whether it serves and promotes cooperation; (2) regulatory constraints/whether it regulates and restricts; (3) accountability/whether it punishes non-cooperative behavior.[7]

3.2.1 Substantive Content/Whether It Serves and Promotes Cooperation

At present, intergovernmental cooperation is still difficult, inefficient, and lacking substantive content. As cybersecurity involves vital interests, countries are active in participating in cybersecurity governance. However, current intergovernmental cybersecurity cooperation only concerns general political fields such as combating cybercrime while cooperation in core military and strategic interests is scarce. Competition and benefit struggle are fierce behind international security cooperation. There is a considerable lack of substantive content in the cybersecurity cooperation between countries as well as between countries and non-state actors.

3.2.2 Regulatory Constraints/Whether It Regulates and Restricts

From the perspective of regulatory constraints, organizations and past international conferences have issued policies and declarations to standardize rules of cyberspace. However, most of these treaties contain general principles, not substantive norms. For example, the Convention on Cybercrime, adopted in 2001, has had a limited role in reducing cybercrime as it has substantive and procedural law provisions to be transposed into domestic law. However, in organizational terms at the technical level cooperation is rather successful: the rules for the distribution of key resources in cyberspace, domain names, and IP addresses have been

[7] He Ye, "Effectiveness Analysis of Global Cybersecurity Governance Mechanism," (Master's thesis, China Foreign Affairs University, 2015), 22–24.

accepted by all countries. Non-governmental platforms such as ICANN, IETF, and IGF dominate the rules of cybersecurity governance and maintain the operation of networks.

3.2.3 Accountability/Whether It Punishes Non-cooperative Behavior

Obviously, due to the virtual nature of cyberspace and the ambiguity of management rights, there is still no international law to punish violations of regulations and non-cooperative behavior. Despite ongoing efforts (see JoonKoo **Yoo** in this volume), the international community has not reached a legally binding document on cybersecurity at the level of the United Nations.

3.3 Confrontation Between Camps

Cybersecurity is a global issue. Countries have common interests in this field but inevitably there are also differences in interests and preferences. Although there should be both cooperation and competition, many areas of cybersecurity are showing a trend of camp-based confrontation. Due to differences in development stages and ideologies, countries are classified into certain camps such as developing and developed countries according to the level of economic and social development. It is relatively straightforward for members in the same camp to reach a consensus based on common interests. At present, international competition surrounding cybersecurity presents a pattern of "one superpower, multiple powers, two camps." "One superpower" refers to the United States whose Internet advantages are clear. "Two camps" refers to two groups of countries that disagree on cybersecurity issues: The first camp consists of developed countries headed by the United States. The second camp consists of developing countries led by China and Russia. This pattern reflects the characteristics of the current international climate, namely antagonism and competition between traditional and emerging powers.

Although alignment is a common phenomenon in international relations, excessive alignment amplifies contradictions, leading to inefficient or even ineffective cybersecurity cooperation.[8] First, it restricts camp members from taking positions based on their own interests and might

[8] Lu Chuanying, "Security Dilemmas Faced by Major Power in Cyberspace: A Case Study of China-Europe Cyber Cooperation," *Chinese Journal of European Studies*, no.2 (2019): 114.

cause a spillover of local differences, which in turn affects global cybersecurity governance. For example, on the issue of the militarization of cyberspace, China and Russia have different views from the United States and the European Union. While the United States and the EU advocate giving countries the right to independently determine and counter cyber attacks with force, China is clearly opposed to the militarization of cyberspace and any provisions that would grant states the lawful use of force in cyberspace. Although it is in the EU's interest to oppose the militarization of cyberspace and to enact provisions against granting states the lawful use of force in cyberspace, the EU finally stands on the side of the United States.[9] As a result, the UNGGE[10] failed to reach a consensus and release a report as scheduled in 2016.[11]

Second, excessive alignment can easily lead to antagonistic policies, constraining countries in different camps from seeking common interests and thereby reducing the space for cooperation. Although countries have diversified policy choices in cyberspace, it is wrong to classify a country as totally adhering to cyber sovereignty at one extreme (in which national governments have greater control over the Internet) or abiding by a multi-stakeholder model as another extreme (in which the Internet is open, accessible, dynamic, and governed collectively by multi-stakeholders). Due to the complexity of cybersecurity issues, no governance model can effectively solve all these problems.[12] While the United States, the UK, Japan, Canada, and Australia advocate an open cyberspace free from interference, the EU emphasizes the application of national sovereignty in cyberspace (such as "digital sovereignty" or "data sovereignty"). China advocates the key role of states in cyberspace governance while also emphasizing multilateral participation under the cyber sovereignty framework. The existence of these camps makes it difficult for countries belonging to a camp to conduct dialogue in accordance with their own positions.

[9] Thomas Renard, "EU cyber partnerships: Assessing the EU strategic partnerships with third countries in the cyber domain," *European Politics and Society* 19, no.3 (2018): 321–337.

[10] UNGGE refers to the UN Group of Governmental Experts on Developments in the Field of Information and Telecommunications in the Context of International Security.

[11] Agreement was reached in 2021. For a discussion on the failure of UNGGE earlier, see also the chapter of **Yoo** Joonkoo in this volume.

[12] Jacqueline Eggenschwiler, "A Typology of Cybersecurity Governance Models," *St Antony's International Review* 13, no. 2 (2018): 64–78.

In short, problems such as fragmented mechanisms, poor effectiveness, and camp-based confrontation are counterproductive for international cybersecurity cooperation. The following section discusses various factors underlying these problems.

4 Reasons for International Cybersecurity Cooperation Dilemma

The problems for international cybersecurity cooperation—as discussed in the previous section—can be explained from the perspective of realism and constructivism. From a realist perspective, differences in interests due to gaps in national power, cost sharing and distribution of benefits, and technical uncertainties are influential factors that make it difficult to achieve international cybersecurity cooperation. From the constructivist point of view, perceptions caused by stereotype thinking and the amplifying effect of the characteristics of cyberspace on misperceptions also impede cooperation.

4.1 Realist Perspective

4.1.1 Different Interests to Cooperate Caused by Gaps in National Power

Like in the fields of economics and military security, countries pursue varying interests in cyberspace. Although every country is aware of the importance of cybersecurity, differences in interests and technological levels cause countries to have varying threat perceptions posed by cyberspace. Therefore, the interest to participate in international mechanisms varies, which renders effective cooperation on a multitude of practical issues difficult. The United States, for example, conceives cyber hackers as a major threat and adopts offensive retaliatory measures against countries that it believes support hackers. For China, although the protection of its digital infrastructure and digital economy are important national interests, the stability of society and politics is its primary consideration.[13] Different

[13] Cai Cuihong, "Cybersecurity in the Chinese Context: Changing Concepts, Vital Interests, and Prospects for Cooperation",*China Quarterly of International Strategic Studies*1, no.3 (2015). For a discussion on China's cyber strategy, see also the chapter of Mason **Richey** in this volume.

interests based on different national priorities are the root cause of the predicament of international cybersecurity cooperation.

4.1.2 *The Game of Cost Allocation and Benefit Distribution*

International cybersecurity cooperation is a collective action. The key to getting rid of the dilemma of collective action lies in how to balance the differences and conflicts between sovereign states and multiple governance entities, how to calculate cooperation costs and benefits as well as to determine what kind of incentive and supervisory system to enact. These all require a reasonable allocation of costs and benefits.[14] The free-rider strategy of countries with weak Internet technologies increases the cost of a cooperative response while reducing their own cost in participation.

In addition to costs being distributed among diverse interest groups, there is also the problem of cost distribution among different approaches. For instance, for parties to existing conventions, the cost of creating a new one is higher than the cost of enlarging the existing one. Countries have not reached a consensus on which approach should be prioritized in a collective response. The possible consequence is that every feasible approach will receive roughly equal (but not dominant) attention, thereby delaying the development of collective security. Countries with advantages in Internet technology are more willing to invest in the application of existing international cybersecurity laws rather than new international conventions. This is because the latter option may provide countries with lower technological capabilities the opportunity to gain discourse power.[15] Technologically disadvantaged countries typically believe that cybersecurity rules should be more than technical parameters: they should also reflect the equality among countries as well as national interest priorities; otherwise, the formulation of relevant cybersecurity rules may inevitably provide a distribution of benefits favorable for countries with advantages in network technology.

4.1.3 *Technical Uncertainties*

Technical factors impede security cooperation in cyberspace in two distinct ways.

[14] Lin Jing, "The Obstacles of International Cooperation Regarding Safeguarding Cybersecurity and China's Strategies," *Journal of Xi'an Jiaotong University (Social Sciences)* 37, no. 2 (March 2017): 80–81.

[15] Ibid., 80–81.

First, there is the factor of technical capacity. Cybersecurity cooperation is different from other non-traditional forms of cooperation as it requires high technological know-how. For example, in responding to cyberterrorism, there are vast differences in terms of country size and economic development on top of an obvious gap in cyber capabilities. If individual states cannot adequately track cyber attacks technologically, their participation in international cybersecurity cooperation is limited and perhaps more words than deeds.

Second, the rapid development of cyber technology outperforms cooperative efforts, as the openness and anonymity of Internet technology increases the difficulty of traceability. In numerous cybersecurity incidents, providing smoking gun evidence to confirm the source of an attack has proved elusive. Therefore, and because of the lack of internationally agreed tracking rules, it is difficult for the international community to take actions to punish alleged attackers.[16] Cybersecurity cooperation mechanisms of some regional organizations and bilateral cooperation mechanisms between governments mainly involve information sharing, education, scientific research, and cybercrime law enforcement. Yet, without adequate technological capabilities, these collaborations are all insufficient to secure cyberspace. The development of the Internet of things (IoT) and artificial intelligence will further increase the difficulty to achieve cybersecurity through cooperation.

4.2 Constructivist Perspective

4.2.1 Perceptions Caused by Stereotype Thinking

Cyberspace is a very different important issue from other topics in international relations. On the one hand, the emergence and development of the Internet has had a significant impact on the international system and national security strategies. Decision makers face a more and more complex environment that is related to cyber. On the other hand, decision makers look at cyberspace through the lens of traditional international relations concepts, existing theories, and knowledge frameworks to understand the behavior and intentions of countries. This often contributes to

[16] Lu Chuanying, "Cyberspace Security Dilemma and the Construction of Governance Institution," 52–53.

misperceptions, an important reason for distrust in the international community impacting negatively on cybersecurity cooperation.

The complex nature of cyberspace requires new theoretical understandings. However, due to the rising complexity and the rapid evolution of this field, the construction of theories lags far behind reality. When there is a lack of novel theories to explain new phenomena, decision makers often resort to utilizing existing theories.[17] A typical phenomenon is to use geopolitical theory to explain international relations in cyberspace. The aforementioned excessive alignment is influenced by traditional geopolitical thinking. Countries classify themselves into specific camps, using the interests of their respective camp to define their own interests, adhering to the mindset that only the governance concepts and methods proposed by their respective camp can meet their interests. Within each camp though, there are also fundamental differences among countries such as between the United States and its allies on the issue of cyberspace militarization.[18] The United States promotes the militarization of cyberspace in pursuit of cyberspace hegemony, which not only creates security risks for its allies but also involves them in a cyber arms race.[19] In addition, the Snowden revelations show that the United States does not trust its allies and also mean that those monitored allies do not trust US national security anymore. This behavior has undermined the national security of its allies.[20] Similarly, although China and Russia agree in their common February 2022 blueprint on the 'International Relations Entering a New Era and the Global Sustainable Development' on the need to reform Internet governance through internationalisation while protecting their sovereignty,[21] the two sides have different perceptions on how to shape the cyberspace governance system. Compared with China's commitment to reform under the existing system, Russia is more inclined to overthrow the current system

[17] Robert Jervis, *Perception and Misperception in International Politics* (Princeton, UK: Princeton University Press, 2017), 143–144.

[18] Wang Congyue, "Inheritance, Changes and Reflections on US-EU Cyber Security Cooperation in the Trump Era," *Foreign Theoretical Trends*, no.7 (2019): 106–116.

[19] George Christou, "Transatlantic Cooperation in Cybersecurity: Converging on Security as Resilience?" in *Cybersecurity in the European Union* (London: Palgrave Macmillan Press, 2016).

[20] "ENISA Threat Landscape Report 2018," ENISA, accessed April 5, 2021, shttps://www.enisa.europa.eu/publications/enisa-threat-landscape-report-2018.

[21] Joint Statement of the Russian Federation and the People's Republic of China on the International Relations Entering a New Era and the Global Sustainable Development, February 4, 2022, http://en.kremlin.ru/supplement/5770

and build anew. Geopolitical thinking causes the two camps to ignore the diverse policy positions of stakeholders, classifying each side into camps. This will reduce mutual trust and impede international cybersecurity cooperation.

4.2.2 Features of Cyberspace Magnifying Perceptions

The disruptive changes that cyberspace brings to international relations and national strategies are challenging the belief systems of decision makers. The advancement of Internet technology and its ever-increasing use continue to blur the boundaries between cyber and physical spaces. New technologies can lead to creative destruction,[22] which allows cyberspace to subvert the physical world's order, norms, and ethics, triggering concerns related to cybersecurity and amplifying false perceptions.

The frequent occurrence of cybersecurity incidents has aggravated cybersecurity concerns and caused many countries to invest more in this realm. Compared to traditional security, cybersecurity has the characteristics of ubiquity, virtuality, and deniability, causing cognitive challenges and aggravating anxiety related to security.[23] The ubiquitous nature of cybersecurity issues means that all devices may have unknown technical security vulnerabilities that may be exploited by adversaries to launch cyber attacks. Similar to traditional military concerns, the virtual nature of cyber weapons makes it impossible for decision makers to have an accurate understanding of the size, level, and degree of harm posed by opponents' cyber weapons, which amplifies the country's sense of insecurity. In addition, the anonymity and transnational nature of the Internet increase the difficulty of tracing the source of a cyber attack. Therefore, an actor who initiates a cyber attack can utilize deniability.[24] In previous cybersecurity incidents, the international community and its stakeholders often have disputes over traceability. It is not easy to determine the identity of an attacker, which makes it hard to apply countermeasures. It is equally difficult for victims to defend their rights, which amplifies decision makers' anxiety about cybersecurity and thus compounds the difficulty of rational

[22] "Creative destruction" is a concept in economics raised by Joseph A. Schumpeter. It describes the "process of industrial mutation that continuously revolutionizes the economic structure from within, incessantly destroying the old one, incessantly creating a new one."

[23] Lu Chuanying, "Security Dilemmas Faced by Major Power in Cyberspace: A Case Study of China-Europe Cyber Cooperation," 122.

[24] Thomas Rid and Ben Buchanan, "Attributing Cyber Attacks," *The Journal of Strategic Studies* 38, no. 1–2 (2015): 4–37.

decision-making. Decision makers tend to exaggerate the cyber threats they face and the strength of their opponents, impeding international cooperation due to a lack of mutual trust.

In addition, new features of cyberspace often act to amplify the security dilemma. The security dilemma is a basic concept in the field of international securitywhich explains the fear of one country for the growth of another's strength. The development of corresponding capabilities to deal with an emerging power results in a situation of vicious competition, possibly even causing an arms race involving traditional military technologies. With the intensification of misperceptions, it may also trigger international wars. American scholar John H. Herz believes that international politics is a game for power and that the pursuit of hegemony is the best way to ensure national security in an anarchic international system.[25] In cyberspace, this kind of anarchy and uncertainty is more evident. In a state of anarchy, a country's unclear information regarding the strategic intentions of others can easily lead it to adopt a comprehensive cybersecurity defense system or even offensive system, leading to competition and confrontation in cybersecurity.

5 A Primer on the Predicament of International Cybersecurity Cooperation in EU-South Korea Relations

Cross-border cooperation in cybersecurity domains is conditioned by numerous political realities. Some of the factors incentivize cooperative behavior between the European Union (EU) and South Korea, while some disincentivizing them.

On the incentive side, the EU and South Korea are both major powers in cyberspace. Both the EU and South Korea are early adopters and utilizers of cyber technology.[26] Both have high Internet penetration rates and advanced Internet technology. The EU's digital technology is in a leading position in the world, ranking high in the field of cutting-edge information technology. They also have a number of key technical standards organizations and information technology industry giants. In addition, South

[25] John H. Herz, "Idealist Internationalism and Security Dilemma," *World Politics*, no. 2 (1950), 157–158.

[26] For a discussion on this, see also the chapter of Michael **Reiterer** in this volume.

Korea has become a "middle power" with resources and international influence sufficient to impact the international system in cyberspace.

As a result of these similarities, the two have common interests and demands in the field of cybersecurity. The possibility of conflict on cost allocation and benefit distribution is correspondingly reduced due to the minimal technological gap, providing material and technical guarantees for cooperation between the two. At the same time, the similarity of their values and ideology reduces the risk of misperception between them and reduces obstacles for their cooperation on cybersecurity.

In addition, despite the fragmentation of international cybersecurity cooperation mechanisms, a somewhat institutionalized cybersecurity cooperation mechanism has been established between the EU and South Korea based on a strategic partnership reflecting shared values and interests. In the dimension of cybersecurity, this strategic partnership is embodied in a Framework Participation Agreement (signed in 2014), which is about South Korea's participation in EU's Common Security and Defence Policy (CSDP) operations, and the EU-ROK Cyber Dialogue, which is an annual meeting of officials since 2013.

Both the EU and South Korea are US allies, but the alliance of US and South Korea exhibits stronger path dependency. In the "Joint Communique of the 52nd U.S.-Republic of Korea Security Consultative Meeting" released on October 14, 2020, the two sides committed to maintain close communication and coordination regarding the cyber domain through sharing trends of cyber threats as well as discussing corresponding policy changes. They also concurred in the need for exchanges between their respective cyber commands with the aim of discussing and promoting mutual interests.[27] Similarly, the EU also has a long history of cybersecurity collaboration with the United States, covering almost all areas such as infrastructure, technical arrangements, data protection, cybercrime prevention as well as joint actions such as the NATO Cyber Coalition Exercise. All these EU-US or ROK-US high-level security and defense cooperation continues to overshadow and inhibit the deepening bilateral engagement between South Korea and the EU in the security domain.

[27] U.S. Department of Defense, "Joint Communique of the 52nd U.S.-Republic of Korea Security Consultative Meeting", accessed July 5, 2021, https://www.defense.gov/Newsroom/Releases/Release/Article/2381879/joint-communique-of-the-52nd-us-republic-of-korea-security-consultative-meeting/.

Lastly, some factors are ambiguous, having both positive and negative effects on EU-ROK cooperation in cybersecurity. South Korea and the EU are both in the US-led camp on cyber issues, especially concerning multi-stakeholder governance. The same bloc identity can contribute to a common position between the EU and South Korea, which is helpful for their talks. But the existence of this camp identity also hinders their autonomy in security cooperation. The strategic competition between China and the United States intensifies the confrontation between camps, pushing other states to hedge between the great powers. The EU, like South Korea, relies on the United States for security and on China for trade. It thus behooves South Korea and the EU to choose their hedging behavior and seek strategic autonomy. Both of them need to look for political, economic, diplomatic, and military partners beyond the United States and China, and should strive to explore new space for cybersecurity cooperation to realize their own strategic autonomy.

6 Conclusion

Present international cybersecurity cooperation has manifested itself in cooperation models involving many participating countries. Despite this, the actual situation is not satisfactory. There are numerous problems including mechanism fragmentation, poor effectiveness of international mechanisms and camp-based confrontation. From a realist point of view, these problems are the result of cost- and benefit-sharing disagreements, limitations due to technical uncertainties, differences in interests as well as discrepancies in terms of willingness to cooperate due to gaps in national power and stage of development. From a constructivist perspective, contributing factors such as misperceptions caused by stereotypes and characteristics of cyberspace amplify false perceptions.

The establishment of an international cybersecurity cooperation mechanism is a dynamic process of continuous learning, evolution, and identification. Faced with the emerging field of cyberspace, which is still in its infancy, countries still need time to gain experience and interact while gradually exploring means of cybersecurity cooperation in practice. When handling the new issues associated with cyberspace, decision makers also need to constantly update their knowledge to better cope with new challenges. Perceptions in the field of cybersecurity should be formed through building rational knowledge frameworks and channels for information exchange in addition to strengthening agenda setting on new issues. The

EU and South Korea can take steps to improve the quality of their cooperation on cybersecurity, while contributing to international cooperation on cybersecurity.

In terms of mechanism fragmentation, there are already institutionalized arrangements for cybersecurity cooperation between the EU and South Korea. But there is still a need to strike a balance between a certain level of systematization of the EU's policies and a customized partnership, because a tailor-made framework is essential to ensure effectiveness. In addition to integrating cybersecurity cooperation into existing security cooperation channels, consolidating existing but parallel cybersecurity cooperation into dedicated cybersecurity cooperation mechanism is of great significance.

In terms of the confrontation between camps, the EU and South Korea need to carefully consider factors which could incentivize or hinder EU-South Korea cybersecurity cooperation because of their cooperation with the United States. This means that the EU and South Korea need to participate in international cybersecurity cooperation from their own objective interests, to gain more space for strategic autonomy and to avoid the negative impact of misperceptions on cooperation.

At a time when major powers are competing in cyberspace, maintaining the basic strategic stability of cyberspace and adopting a cautious attitude toward cybersecurity cooperation are not only in the interests of the EU and South Korea but also in the common interest of stakeholders across the world.

References

Cai, Cuihong. "Cybersecurity in the Chinese Context: Changing Concepts, Vital Interests, and Prospects for Cooperation", *China Quarterly of International Strategic Studies* 1, no. 3 (2015).

Christou, George. "Transatlantic Cooperation in Cybersecurity: Converging on Security as Resilience?" in *Cybersecurity in the European Union*, London: Palgrave Macmillan Press, 2016.

Donahoe, Eileen. "Governance Challenges in the Global Digital Ecosystem," In *Innovations in Global Governance: Peace-Building, Human Rights, Internet Governance and Cybersecurity, and Climate Change, Council on Foreign Relations*, 24–28. https://www.cfr.org/report/innovations-global-governance (accessed April 5, 2021).

Eggenschwiler, Jacqueline. "A Typology of Cybersecurity Governance Models," *St Antony's International Review* 13, no. 2 (2018): 64–78.

Herz, John H. "Idealist Internationalism and Security Dilemma," *World Politics*, no. 2 (1950): 157–158.

He Ye. "Effectiveness Analysis of Global Cybersecurity Governance Mechanism" (Master's thesis, China Foreign Affairs University, 2015), 22–24.

Jervis, Robert. *Perception and Misperception in International Politics* (Princeton, UK: Princeton University Press, 2017), 143–144.

Lin Jing. "The Obstacles of International Cooperation Regarding Safeguarding Cybersecurity and China's Strategies," *Journal of Xi'an Jiaotong University (Social Sciences)* 37, no. 2 (March 2017): 80–81.

Lu Chuanying. "Cyberspace Security Dilemma and the Construction of Governance Institution," *Contemporary International Relations*, no. 11 (2018): 52–53.

OECD. *OECD Guidelines for the Security of Information Systems and Networks: Towards a Culture of Security*, OECD.org. https://www.oecd.org/digital/ieconomy/oecdguidelinesforthesecurityofinformationsystemsandnetworkstowardsacultureofsecurity.htm (accessed June 25, 2021).

Renard, Thomas. "EU cyber partnerships: Assessing the EU strategic partnerships with third countries in the cyber domain," *European Politics and Society* 19, no. 3 (2018): 321–337.

Rid, Thomas, and Ben Buchanan. "Attributing Cyber Attacks," *The Journal of Strategic Studies*, 38, 2015, no. 1–2 (2015): 4–37.

Sofaer, Abraham D., and Seymour E. Goodman. *Proposal for an International Convention on Cyber Crime and Terrorism*, CISAC. https://cisac.fsi.stanford.edu/publications/proposal_for_an_international_convention_on_cyber_crime_and_terrorism_a (accessed June 25, 2021).

Tardy, Thierry. "Revisiting the EU's Security Partnerships," European Union Institute for Security Studies (EUISS). https://www.iss.europa.eu/content/revisiting-eu%E2%80%99s-security-partnerships (accessed April 5, 2021).

U.S. Department of Defense. "Joint Communique of the 52nd U.S.-Republic of Korea Security Consultative Meeting". https://www.defense.gov/Newsroom/Releases/Release/Article/2381879/joint-communique-of-the-52nd-us-republic-of-korea-security-consultative-meeting/ (accessed July 5, 2021).

Wang Congyue. "Inheritance, Changes and Reflections on US-EU Cyber Security Cooperation in the Trump Era," *Foreign Theoretical Trends*, no. 7 (2019): 106–116.

Conclusion: From Words to Deeds

Michael Reiterer, Ramon Pacheco Pardo, and Gertjan Boulet

1 Key Rationales for Cybersecurity Cooperation Between the EU and South Korea

Cybersecurity cooperation between the EU and South Korea ought to move from theory to practice, from words to deeds. This request is strengthened by what appears to be anecdotal evidence: The 2021 Financial Times, McKinsey Award for the Business Book of the Year went

---

M. Reiterer (✉)
Centre for Security, Diplomacy & Strategy (CSDS), Brussels School of Governance (BSoG-VUB), Brussels, Belgium
e-mail: michael.reiterer@vub.be

R. Pacheco Pardo
King's College, London, UK
e-mail: Ramon.Pacheco.Pardo@vub.be

G. Boulet
Korea University, Seoul, South Korea

Vrije Universiteit Brussel (VUB), Brussels, Belgium
e-mail: gertjan.boulet@vub.be

© The Author(s), under exclusive license to Springer Nature Switzerland AG 2022

G. Boulet et al. (eds.), *Cybersecurity Policy in the EU and South Korea from Consultation to Action*, New Security Challenges,
https://doi.org/10.1007/978-3-031-08384-6_14

to *This Is How They Tell Me the World Ends* by Nicole Perlroth,[1] praised as an eye-opener to the potential of a cyber catastrophe and a call to get the issue on the agenda of CEOs and policymakers.[2] The same *FT* edition carries a comment arguing that "Rules of war need rewriting for the AI era,"[3] another wake-up call. The Russian aggression of Ukraine included cyber attacks and provided more proof than was needed!

This meets the rationale for this volume: as spelt out by the co-editors in the Introduction, techno-competition in the digital economy boosted by the covid-19 crisis, the growth of critical infrastructure and the need for its protection against cyber threats, and the growing weaponization of technology have turned technological competition into an arms race. All these elements require joint endeavours of like-minded partners such as the EU and South Korea to push for cyber governance as part of a rules-based system. This can contribute to the much-needed confidence-building with trust in short supply as evidenced by the difficulties in finding common ground in the United Nations when discussing the applicability of international law to the cyber domain (JoonKoo **Yoo**, Mason **Richey**).

The authors of this volume have taken up these burning issues and argue that the EU and South Korea are particularly well placed to cooperate to increase their cybersecurity and resilience as they pursue similar goals. Therefore, they have an interest in aligning their strategies and policies. Technical capabilities paired with shared values and a belief in the need for multilateralism in this field provide a strong basis to meet the often same types of malicious cyber activities by the same (non-)state actors operating in the cyber domain. Geopolitics provides a further incentive for cooperation: in the Sino-US competition both the EU and South Korea are interested to cooperate with the US to face up to China's growing power and techno-nationalism, but would like to pursue these goals "their way" which could become a common EU-Korean way.

[1] Nicole Perlroth, *This Is How They Tell Me the World Ends*. Bloomsbury Publishing, 2021.

[2] Andrew Hill, *Sobering probe of cyber weapons wins business book of the year*. Financial Times, 2 December 2021; p. 2.

[3] Financial Times, 2 December 2021; p. 18.

2 Cybersecurity Policy Development in the EU and South Korea

The EU has been prioritizing cybersecurity dating back to the early 2010s. Most notably, Brussels launched its first Cybersecurity Strategy in 2013. This strategy was supplemented and took a clearer shape by the inclusion of cybersecurity in the European Commission's European Agenda on Security 2015–2020. This inclusion in the Agenda broke down the no longer tenable differentiation between internal and external security. The Strategy was also reinforced with the Directive on security of Network and Information Systems (the NIS Directive of 2016), the first-ever EU-wide security law. In 2019, the Strategy was further strengthened when the EU Cybersecurity Act entered into force.[4] The Strategy and legal instruments have served to reinforce the mandate of the European Union Agency for Cybersecurity (ENISA), operational since 2005 but whose remit has expanded significantly in recent years.[5]

Many authors (Mason **Richey**, Joohee **Park**, Anna-Maria **Osula**, Tatiana **Tropina**, and Gibum **Kim**) deal with the criminal energy spent on developing Wannacry, NotPetya, Fastcash Campaign, Operation Cloud, attacks on Sony, the Bangladesh Bank Heist, and the Digital Currency Exchange Hack. These attacks brought hybrid attacks to new heights and forced policymakers to speed up cooperation in an area which used to rely on interstate cooperation based on traditional rules and customs, which are slow, cumbersome, and no longer adequate. This is also testimony that the times where domestic and international security could be neatly separated are gone and that a holistic view is the only one which can bring solutions in the cyber world. As the case of North Korea shows, ransom business can be turned into an important budgetary item. The description of the North Korean system by Kyu-dok **Hong** and Seong-jong **Song** is revealing in this respect.

In 2020, the European Commission and the High Representative of the Union for Foreign Affairs and Security Policy launched a new EU Cybersecurity Strategy. This shows the importance that the EU confers on

[4] European Commission, *New EU Cybersecurity Strategy and New Rules to Make Physical and Digital Critical Entities More Resilient*, 16 December 2020, available at https://ec.europa.eu/commission/presscorner/detail/en/ip_20_2391 (accessed 26 November 2021).

[5] European Union Agency for Cybersecurity (ENISA), *ENISA Mandate and Regulatory Framework*, 2021, available at https://www.enisa.europa.eu/about-enisa/regulatory-framework (accessed 26 November 2021).

this area of its own security. The Strategy emphasizes three areas of action for the coming years: resilience, technological sovereignty and leadership; building operational capacity to prevent, deter, and respond; and advancing a global and open cyberspace through increased cooperation.[6] Recognizing that cyber is a prime area for multilateral cooperation because of its very nature, the EU strives with its cyber diplomacy to realize an open cyberspace. To this end it developed, as Michael **Reiterer** presents in this volume, its own cyber diplomacy toolbox which gives the EU autonomous tools while inviting partners like South Korea to join hands, as autonomy does not equal cyber nationalism.

South Korea pursued a similar path when developing its cybersecurity policy. In 2009, the Lee Myung-bak government released the Comprehensive Countermeasures for National Cyber Crisis. This was followed by the Masterplan for National Cybersecurity in 2011. The Park Geun-hye government came into office in 2013, and within a few months issued the National Cybersecurity Measure. These documents, however, created a patchwork of measures rather than a well-defined, holistic strategy, and focused heavily on the North Korean cyber threat.[7] It was the Moon Jae-in administration that took a more comprehensive and strategic approach in 2019 with the National Cybersecurity Strategy (NCS).[8]

This 2019 NCS emphasizes three key elements: a focus on a broad set of threats including those coming from other countries; emphasis on deterrence including through civilian-government-military cooperation; and a clear commitment to global cooperation, including the development of global norms and capacity building in developing countries.[9] Thus, like the EU, South Korea strives to develop its own capabilities while remaining open for international cooperation.

[6] European Commission and High Representative of the Union for Foreign Affairs and Security Policy, *The EU's Cybersecurity Strategy for the Digital Decade*, 16 December 2020, available at https://ec.europa.eu/newsroom/dae/document.cfm?doc_id=72164 (accessed 26 November 2020).

[7] Ramon Pacheco Pardo, Tongfi Kim, Maximilian Ernst, Sung Kyoo Ahn and Riccardo Villa, *Beyond Traditional Security: South Korea's Positioning towards the Cyber, Energy, Maritime and Trade Security Domains* (Brussels: KF-VUB Korea Chair, 2020), p. 10.

[8] National Security Office of Cheong Wa Dae, *National Cybersecurity Strategy*, April 2019, available at https://www.itu.int/en/ITU-D/Cybersecurity/Documents/National_Strategies_Repository/National%20Cybersecurity%20Strategy_South%20Korea.pdf (accessed 26 November 2021).

[9] Ramon Pacheco Pardo et al., *Beyond Traditional Security*, pp. 10–11.

In consideration of these similar approaches, the EU and South Korea have launched a bilateral Cyber Dialogue in 2014. At the last session of this dialogue (November 2020) at the time of writing, both sides confirmed their commitment "to an open, free, stable and secure cyberspace."[10] In 2020 the dialogue was enlarged to a Track 1.5 EU-Republic of Korea Cyber Consultations.[11] Despite these activities and the EU choosing South Korea as the partner for a pilot project on cyber cooperation in implementing its 2018 Enhancing Security Cooperation in and with Asia project, the authors of this volume agree on the need to move from dialogue to action.

3 Towards EU-South Korea Cybersecurity Cooperation Through Action

Based on the common goal of ensuring an open cyberspace—as opposed to a government-controlled space—and the similarity of threats faced by the EU and South Korea, a common agenda for cooperation between South Korea and the EU can be developed, as the contributions to this volume show.

According to Michael **Reiterer** the EU considers South Korea on top of the Strategic Partnership as one of its key partners in the area of cybersecurity. Thus, it has established a framework to boost bilateral security ties. In 2020, Brussels launched the programme Enhancing Security Cooperation in and with Asia (ESIWA). Initially running until 2024, ESIWA has the mandate to strengthen practical cooperation between the EU and Asia, with eight pilot countries as a starting point. South Korea has been chosen as the focal point for cybersecurity cooperation.[12] This commitment has been enhanced when the Commission and the High Representative announced the EU's Indo-Pacific strategy (IPS) in September 2021. The IPS includes specific actions to boost cooperation in the cyber domain with strategic partners, such as South Korea. Actions include negotiating a Digital Partnership Agreement that could serve to

[10] European Commission, *6th Annual European Union – Republic of Korea Cyber Dialogue*, 25 November 2020, available at https://cdn4.euraxess.org/worldwide/south-korea/6th-annual-european-union---republic-korea-cyber-dialogue (accessed 26 November 2021).

[11] EU Cyber Direct, *EU-ROK Cyber Consultations*, 8 October 2020, available at https://eucyberdirect.eu/events/eu-rok-cyber-consultations (accessed 26 November 2021).

[12] GIZ, *Enhancing Security Cooperation in and With Asia*, 2021, available at https://www.giz.de/en/worldwide/87412.html (accessed 26 November 2021).

develop joint standards, conclude the data protection adequacy negotiation process (on 17 December 2021, the EC adopted an adequacy decision for South Korea), and establish joint R & D under the aegis of Horizon Europe.[13] Cooperation within this framework could also boost links between both partners and other countries and institutions in the Indo-Pacific including ASEAN, India, or Japan.[14] Network diplomacy is particularly important and effective in the cyber area.

George **Christou** and Ji Soo **Lee** suggest data protection, cybersecurity, and emerging technologies as areas in which the EU and South Korea could intensify their cooperation. In spite of lingering differences in their approaches, there is clear overlap of interests on which they can build. In the area of data protection, the EU approach of striving for adequacy and not harmonization facilitated South Korea's decision to sign up for the EU system which was in turn an important step in globalizing the GDPR system. Emerging technologies require technical cooperation, joint research, and the development of compatible certification schemes.

In order to successfully provide cybersecurity in creating synergies, compatible approaches, and frameworks, the EU and South Korea have to ensure that their various agencies involved in cybersecurity cooperate. On cyber resilience, George **Christou** and Ji Soo **Lee** suggest that Brussels and Seoul can boost information sharing, engage in joint exercises and scenario planning, and increase the number of joint research projects. For cyber conflict prevention, the EU and South Korea can coordinate capacity building initiatives as part of confidence-building measures. This is supported by JoonKoo **Yoo** advocating joint confidence-building measures (CBMs) in the respective regional context of the Organisation for Security and Cooperation in Europe (OSCE) and the ASEAN Regional Forum (ARF).

On emerging technologies, this requires technical cooperation, joint research, and the development of compatible certification schemes. As both the EU and South Korea strive to assure a human-centric digitalization, the launch of a high-level dialogue on the digital economy will facilitate such a comprehensive approach, including research and innovation to

[13] European Commission and High Representative of the Union for Foreign Affairs and Security Policy, *The EU Strategy for Cooperation in the Indo-Pacific*, 16 September 2021), pp. 10–11, available at https://eeas.europa.eu/sites/default/files/jointcommunication_2021_24_1_en.pdf (accessed 26 November 2020).

[14] Linde Desmaele, Maximilian Ernst, Tongfi Kim, Ramon Pacheco Pardo and Michael Reiter, *The EU's Indo-Pacific Strategy: Prospects for Cooperation with South Korea* (Brussels: KF-VUB Korea Chair, 2021), p. 31.

master the challenges of the Fourth Industrial Revolution and the ethical implementation of AI technologies.

On cooperation, Mason **Richey**, in addition to providing a valuable "Cyber Threat Overview," underlines two aspects: the EU's experience in conceptualizing and implementing "regional confidence-building measures in the security arena" is of value not only to South Korea but also to East Asia with respect to cyber threat-reduction and conflict prevention. Secondly, norm-building in the cyber domain in drawing on the experience of the private sector in particular through the setting of norms for hardware standards would enhance the security of networks. Furthermore, as not all intrusions in the cyber realm are automatically illegal, with borderlines between espionage, information collection, and a cyber attack, joint EU-South Korea work in drawing on their technological expertise would be beneficial. He joins others (JoonKoo **Yoo,** Cuihong **Cai**) in suggesting that such a cooperation could lead to minilateral initiatives which could create momentum for the larger players to engage in rule-making.

On cyber sanctions, in the context of the divide between democracies and autocracies in the area of cybersecurity, democracies have woken up to the need to assure the security of their critical infrastructure, including securing the heart of democracy, elections, and given themselves a toolkit which includes sanctions. As Joohee **Park** indicates, the EU launched a legislative act to impose cyber sanctions on third parties in May 2019.[15] In July 2020 the EU imposed its first-ever cyber sanctions, targeting China, North Korea, and Russia.[16] Joohee **Park** explains that South Korea does not have such a framework. She comes to the same conclusion as Michael **Reiterer** that Seoul may be willing to join the EU and other actors such as the US by imposing cyber sanctions jointly. A multilateral framework would increase effectiveness and open the door to defensive actions for countries like South Korea, which are hesitant for geopolitical reasons to act alone. Thus, there is scope for Brussels to share its experience with

[15] Council of the European Union, *Council Decision Concerning Restrictive Measures Against Cyber-Attacks Threatening the Union and its Member States*, 14 May 2019, available at https://data.consilium.europa.eu/doc/document/ST-7299-2019-INIT/en/pdf (accessed 26 November 2021).

[16] Council of the European Union, *EU Imposes the First-Ever Sanctions Against Cyber-Attacks*, 30 July 2020, available at https://www.consilium.europa.eu/en/press/press-releases/2020/07/30/eu-imposes-the-first-ever-sanctions-against-cyber-attacks/ (accessed 26 November 2020).

Seoul and provide support if the South Korean government decides to establish its own cyber sanctions framework.

On cyber arms control, Sangho **Lee** and Maximilian **Ernst**, on the one hand, and Michael **Zinkanell**, on the other, focus on arms control in South Korea and the EU. Sangho **Lee** and Maximilian **Ernst** zoom in into the threat posed by North Korea, including cybercrime and terrorism. As a long shot they suggest that the solution may lie in an inter-Korean cyber arms control agreement. More realistically, at least in the short term, Sangho **Lee** and Maximilian **Ernst** have analysed South Korea's protection against malicious cyber activities originating in North Korea as well as China. They suggest that a combination of intelligence gathering and intelligence-sharing with like-minded partners, joint work to anticipate and mitigate the effects of cyber attacks, and measures to ensure that the population is resilient against phychological cyber warfare (e.g., misinformation campaigns) is all necessary.[17]

From an arms control perspective, Michael **Zinkanell** identifies several ways in which the EU has sought to protect itself against cyber attacks. They include enforcing standards and regulations to ensure that harmful programmes do not have the desired effects, restricting malicious behaviour, and making use of sanctions (see more below). The author believes that these measures have helped to limit the effectiveness of cyber attacks. Therefore, an arms control framework seems appropriate to boost cybersecurity. As he does not expect any international agreement on the use of AI any time soon, he suggests the EU taking measures to safeguard its AI infrastructure, to assure human oversight for its use particularly in the military field, and to set standards as a catalyst for the international community to follow the example.

The work by Sangho **Lee** and Maximilian **Erns**t and the chapter by Michael **Zinkanell** suggest that the EU and South Korea are not necessarily following the same approaches when it comes to cybersecurity, at least when considered from an arms control perspective. This would open the door for cooperation. South Korea could share its experience with intelligence gathering and sharing, if not with the EU at least with some member states. And in fact, the launch of the Five Eyes Plus framework in late 2019 focusing on the North Korean threat scenario suggests that this

[17] Maximilian Ernst and Sangho Lee, 'Countering Cyber Asymmetry on the Korean Peninsula: South Korea's Defense against Cyber Attacks from Authoritarian Sates', *Journal for Intelligence, Propaganda and Security Studies*, Vol. 15, No. 1 (2021): pp. 165–179.

is already happening; France is part of this group.[18] Seoul's incipient focus on resilience should also offer valuable lessons for the EU. In turn, South Korea would profit from the EU's focus on standards and regulations.

Focusing on cyber defence and cyber resilience, Kyu-dok **Hong** and Seong-jong **Song** look at the case of South Korea while Anna-Maria **Osula** analyses the EU. Kyu-dok **Hong** and Seong-jong **Song** suggest that South Korea's capabilities in this area are still limited. They point out that South Korea has a decentralized structure that weakens cybersecurity, with the Ministry of National Defense, the National Intelligence Service, and the Ministry of Security and Public Administration all having their own cybersecurity units. According to the authors, this should give way to a centralized command structure, a whole-of-the-government-approach, consolidating inter-agency cooperation. In addition, the authors point out that South Korea should take a "defend forward" approach whereby South Korea proactively and consistently engages with its adversaries, so that the country can withstand cyber attacks. Therefore, they see adding cyberspace to the coverage of the ROK-US military alliance as an important step. Cuihong **Cai** disagrees with such an approach. Seen against the camp character of the Sino-US competition, her advice is for the EU and South Korea "to participate in international cybersecurity cooperation from their own objective interests, to gain more space for strategic autonomy and to avoid the negative impact of misperceptions on cooperation," thereby introducing the "strategic autonomy" discussion.

Anna-Maria **Osula** focuses on cyber resilience, which, as explained above, is a key element of the EU's 2020 Cybersecurity Strategy. While she regards the EU's existing regulatory framework as a base, she advises the EU to develop pan-EU cyber defence capabilities. In this respect, it is crucial for the EU to advance cutting-edge technologies and the operational capability necessary to prevent, deter, and respond to cyber attacks. The author also underlines the need to cooperate with partners.

Since the authors dealing with cyber defence and resilience suggest that neither South Korea nor the EU already have well-developed strategies, they should join brains and establish cooperation mechanisms. Authors draw attention to the difficulty of democracies in general, and therefore also for the EU and South Korea, to implement highly effective defence

[18] Jagannath Panda, 'Is Seoul Prepared to Join a Five Eyes Plus Framework', *38 North*, 24 August 2020, available at https://www.38north.org/2020/08/jpanda082420/ (accessed 26 November 2021).

mechanisms against cyber attacks because of restraints posed by fundamental rights. Therefore, they have been less assertive than autocracies. Nevertheless, they need to boost their deterrence and response capabilities to protect these rights. Centralization of cyber defence command capabilities is an area for exchange of best practices. What appears as an organizational matter has a strong human rights element because of the need to exercise democratic control of powerful cyber authorities; in this context, South Korea and the EU can learn from each other.

Concerning cybercrime, the EU and South Korea have different legislative frameworks. However, they can engage in information sharing, consultations to make legislation compatible with the Budapest Convention (established by the Council of Europe and entered into force in 2004), exchanging best practices, and developing multi-stakeholder initiatives.

The widely differing views on the Convention and its further development are also reflected by the authors of this volume. This can be exemplified by the treatment of digital evidence regulated in the Second Additional Protocol to the Budapest Convention (as approved by the Committee of Ministers on 17 November 2021[19]). While Tatiana **Tropina** opposes direct access to e-evidence from service providers, Gibum **Kim** is in favour. Tatiana **Tropina** takes a cautious approach highlighting that "there is a need to employ more scrutiny for the mechanisms based on direct requests to providers and, ultimately, to improve existing mutual legal assistance frameworks rather than circumvent them." She sees a trend to this end as "several legislative solutions emerged in recent years, notably: the US CLOUD (The Clarifying Lawful Overseas Use of Data) Act 2018, the EU E-Evidence proposals 2018, and the Second Additional Protocol to the Budapest Convention. To a greater or lesser extent, these frameworks aim to replace the current mutual legal assistance mechanisms with direct requests to providers located abroad." She also sees a need for the EU to improve its current mechanism in aligning it with global efforts based on the Second Additional Protocol while preserving a high level of safeguards for future solutions that will improve access to e-evidence without compromising on fundamental rights.

[19] Council of Europe, *Second Additional Protocol to the Convention on Cybercrime on enhanced co-operation and disclosure of electronic evidence*. 7 November 2021, at https://www.coe.int/en/web/cybercrime/t-cy-drafting-group.

JoonKoo **Yoo** and Cuihong **Cai** agree on the desirability of international norms in order to reduce ambiguity around the applicability of international law and norms. They see the UN at the centre of this endeavour but have to recognize that the societal split we observe nationally also divides the international community. Cuihong **Cai** refers to "camps" which approach the problem from very different angles: either applying existing international law or concluding a new treaty. Keeping cyberspace secure and safe gets different connotations: secure for transactions or safe contentwise. In both cases "openness" has a quite different meaning: functionality vs. ideology. Whether technology is seen as "neutral" or "good"—a value judgement with the yardstick in the eye of the beholder—reflects this split. Considering the unlikeliness of finding common ground in the foreseeable future JoonKoo **Yoo** suggests the EU taking the lead in advancing cyber governance, to create momentum for a broader rules-based framework, with the OSCE and ARF as regional stepping stones to achieve a global solution.

For Cuihong **Cai**, regional cooperation is one of the three possible models of intergovernmental cooperation in addition to global and bilateral. Although her outlook on cooperation between the EU and South Korea based on shared interests is rather positive, one of the main obstacles inhibiting the deepening engagement between South Korea and EU in the security domain in her view is the respective "high level security and defense cooperation" with the US, where South Korea shows a stronger path dependency than the EU. "The same bloc identity can contribute to a common position between the EU and South Korea, which is helpful for their talks. But the existence of this camp identity also hinders their autonomy in security cooperation." Recalling the trade dependency of the EU and South Korea on China, Cuihong **Cai**'s advice is clear: "Both, the EU and South Korea, need to look for political, economic, diplomatic, and military partners beyond the US and China, and should strive to explore new space for cybersecurity cooperation to realize their own strategic autonomy."

4 The EU, South Korea, and Multilateral Cybersecurity Cooperation

Recent EU policy papers, the 2020 Indo-Pacific Strategy as well as the extension of the 2018 Connectivity Strategy, the 2021 Global Gateway,[20] emphasize the importance of cybersecurity and the need for multilateral approaches, thereby widening the scope of the EU cyber diplomacy. As contextualized by Michael **Reiterer**, this diplomacy fulfils three important functions: first, supporting externally the relevant internal EU policies; thereby providing, secondly, a protective as well as, thirdly, an offensive tool through the EU's Cyber Diplomacy Toolbox. The need to include the private sector in these initiatives to be effective has to be underscored as big tech companies have become de facto norm-setters in the field of digital policy.

EU-South Korea cybersecurity cooperation cannot happen in a vacuum. Unfortunately, UN efforts to develop universal cyber rules and norms have stalled.[21] Countries such as China and Russia have an approach to cyberspace that is anathema to the EU, South Korea, and like-minded partners such as Australia, Canada, Japan, or the US. Cuihong **Cai** explains in her contribution the divergent views of what she calls "two camps." While it does not appear likely that a universal framework to govern cyberspace in the short- or even mid-term will be reached, Cuihong **Cai** remains optimistic that there is a common interest to maintain a "basic strategic security of cyberspace." George **Christou** and Ji Soo **Lee** second these views:

> Notwithstanding their own specific regional cyber threats, the EU and South Korea share common interests that will enhance prospects for future collaboration and cooperation in order to achieve shared normative digital goals. The broader challenge for the EU and South Korea will be to ensure not just that other like-minded states remain engaged in achieving a secure, safe, open and free Internet, but that compromises can be found with those states that have alternative visions that can potentially undermine such a vision if consensual norms for state and non-state actor behaviour in cyberspace are not implemented beyond discursive outcome.

[20] Joint Communication to the European Parliament, the Council, the European Economic and Social Committee, the Committee of the Regions and the European Investment Bank (2021). *The Global Gateway*. JOIN (2021) 30 final, 1 December 2021; at https://ec.europa.eu/info/sites/default/files/joint_communication_global_gateway.pdf.

[21] Anders Henriksen, 'The End of the Road for the UN GGE Process: The Future Regulation of Cyberspace', *Journal of Cybersecurity*, Vol. 5, No. 1 (2019), pp. 1–9.

Mason **Richey** rightly warns that a "lack of norms incentivizes predatory behavior" leading to a sort of cancer undermining cybersecurity to a degree that the elimination of cyber threats or attacks is realistically no longer possible. Therefore, it remains to "mitigate cyber escalation risk." Out of three "baskets," "Seoul and Brussels could potentially serve as a core for a "minilateral" caucus (including other East Asian states) providing momentum for major cyber players to increase investment in improved international cybersecurity governance."

Mason **Richey**'s observation that "International law is currently too inchoate to solve many international cybersecurity vulnerabilities, so until that regulatory environment matures, norm-construction will (a) have to suffice and (b) require cooperative middle power actors [like the EU and South Korea] as proponents" leads to JoonKoo **Yoo**.

In drawing on his experience of working within the UN system, JoonKoo **Yoo** combines pessimism and optimism: The latter is fuelled by the adoption of the final OEWG report on 12 March 2021. "The second OEWG (2021–2025) based on a Russian proposal, will have to navigate intense competition between the two blocs seeking to wrest control. The new OEWG mandate will determine the future course of the two parallel UN cyber discussion processes. If the OEWG becomes the key forum for cybersecurity deliberation that does substantive work, discussions on cybersecurity at the UN could be brought back on track." This could eventually lead to a single structure of discussion at the UN. His discussion of "technological neutrality" vs "technology for good" (China) while states are navigating the "capability-vulnerability paradox" is telling.

Brussels and Seoul should therefore be part of and even lead efforts for like-minded partners to cooperate in the area of cybersecurity. Both of them have a high level of digitization, especially South Korea. And both of them have a high degree of normative power, especially the EU. They are thus in a position to help advance multilateral cooperation and attempt to build bridges with China.

One potential venue for cooperation is the Budapest Convention. As explained above, the convention is far from being universal. In fact, only 65 parties have joined as of 2021. And since it was drawn up by the Council of Europe rather than by the UN and most countries outside of the Council were excluded from the drafting process, it is unlikely to become universal.[22] Indeed, South Korea is among the many countries

[22] Clough, 'A World of Difference'.

who have originally declined to join the convention. As Gibum **Kim** points out in his chapter, it would be in the interest of South Korea to join the convention. A joint investigation working group with the EU could add teeth to the Convention. Especially since EU member states, Australia, Canada, Japan, and the US are parties to the convention.[23] In other words, South Korea's like-minded partners with which Seoul is engaged in cooperation in the area of cybersecurity are already part of the convention. Therefore, by joining the convention, South Korea would be part of a multilateral framework underpinning the activities it already conducts. In terms of multilateral cooperation, he suggests that the EU and South Korea jointly work on capability building in developing countries.

Cybersecurity governance is fragmented, domestically and globally. Despite the Convention's aim at harmonizing legislation and rules among its members and fostering international cooperation, it has failed to become universal, with weak regulatory harmonization among its signatories.[24] Therefore, it makes sense for the EU and South Korea to cooperate on a bilateral basis while the EU continues pursuing its goal to attract more like-minded partners to sign on, including South Korea which has to finish its accession process. Such a bilateral cooperation between the EU and South Korea as like-minded partners with similar (cyber) values can be extended to the networks and mini- or plurilateral settings in which they are participating.

This view is shared by the leaders of the EU and South Korea at the June 2020 virtual leaders' summit: they "reiterated their resolve to uphold the rules-based international order and looked forward to practical cooperation in areas such as cyber security and countering disinformation and other hybrid threats."[25]

In addition, the EU and South Korea can cooperate in minilateral settings involving the two of them, EU member states, and other like-minded

[23] Council of Europe, *Charts of Signatures and Ratifications of Treaty 185*, 2021, available at https://www.coe.int/en/web/conventions/full-list?module=signatures-by-treaty&treatynum=185 (accessed 26 November 2021).

[24] Jonathan Clough, 'A World of Difference: The Budapest Convention on Cybercrime and the Challenges of Harmonisation', *Monash University Law Review*, Vol. 40, No. 3 (2014): pp. 698–736.

[25] Council of the European Union, *Joint press release: Republic of Korea – EU Leaders' video conference meeting*, 20 June 2020; at https://www.consilium.europa.eu/en/press/press-releases/2020/06/30/joint-press-release-republic-of-korea-eu-leaders-video-conference-meeting/.

partners. Activities such as information sharing, table-top exercises, or capacity building in their countries are more cost-effective if conducted within minilateral forums. Coordination among participants may be more complex, but the results are arguably better as there are more actors with different experiences present at the forum. In this context, Mason **Richey**'s chapter concludes with reflections on how international institutions (at bilateral, regional, and global levels) can improve cybersecurity, notably through the promotion of a regime of information exchange and best practices establishing expectations of (and limits on) tolerable cyber behaviour.

The covid-19 pandemic has brought online more services and boosted further e-commerce, the amount of data processed has skyrocketed, and thereby the vulnerability of all has been increased. The interfaces of economy, technology and national security or privacy, business interests and competitiveness invite more multilateral cooperation which has to be as comprehensive as possible as cyber is open to all, whether like-minded or not. Furthermore, the role of the private sector in norm setting, enhancing technology as well as security needs to be taken into account (cf. Michael **Reiterer** on the Paris Call). The research component and the often pivotal role of start-ups which need an accommodating eco-environment need particular attention. The speed of bringing innovation to the market, thus valorizing the achievements of the innovative community of researchers and research institutions, is a determining factor for success.

Similarly, the interplay between international organizations, such as the International Telecommunication Union for technical matters and the WTO to better regulate e-commerce, determines governance in this sector and needs guidance. The applicability of present and future international law, especially in relation to emerging technologies such as automated weapons systems based on AI and the challenges they also pose in the ethical domain, needs leadership. The human-centred approach of the EU and South Korea makes them apt to play a leading role in these endeavours. As data is the lifeline of the nascent economy, leadership in this highly complex and competitive area of cyber will also grant economic and societal advantages.

George **Christou** and Ji Soo **Lee** put it succinctly: "The EU and South Korea recognise the need to address cybersecurity challenges going forward, with the covid-19 pandemic adding to the number of issues that increased digitalisation brings with it in terms of online privacy, data protection, safety and security." Thus, the chapters of this volume lay the groundwork for the EU and South Korea to establish a bilateral cooperation mechanism and work to establish a multilateral one. Cyber threats are not circumscribed to a single country. The cybersecurity strategies of the EU and South Korea are fairly similar. Both are still enhancing their technical and societal capabilities and related infrastructures implementing their relatively recent cybersecurity strategies. Technological progress dictates time and speed; therefore, for Brussels and Seoul time has come to move from concepts to projects, form words to deeds.

Index[1]

A
Adequacy decision, 4, 43, 52, 58, 59, 192n49, 244
Advanced Persistent Threat (APT), 72, 82, 124, 127, 255
Arms control, xi, 5, 7, 119–133, 137–151, 166, 276
Artificial intelligence (AI), ix, x, 3, 31, 42, 49, 53, 60, 68, 77, 139–141, 149–151, 150n35, 170, 180, 184, 187–190, 188n37, 189n38, 192, 209, 244, 261, 270, 275, 276, 283
Asia, 3, 8, 19, 27–30, 56, 56n67, 70, 157, 158, 168, 169, 191, 192, 243, 246, 273
Asymmetry, 119–133
Attribution, 21, 70, 77–79, 90, 100–103, 109, 109n48, 110, 110n49, 126, 131, 149, 162, 179, 247

B
Blockchain, 72, 180, 188
Budapest convention, 4, 56, 101, 102, 102n10, 173, 190, 203n28, 217–219, 221, 226–227, 227n50, 229, 245, 278, 281

C
Capacity building, 20, 32, 49, 54, 56, 60, 88, 145, 169, 190–192, 200, 203–204, 208, 210, 236, 240, 240n7, 246, 248, 272, 274, 283
CERT, 55, 182, 186, 246
Certification, 46n23, 145, 145n20, 146, 184, 274
Chemical, 27, 128, 130, 131, 165

[1] Note: Page numbers followed by 'n' refer to notes.

© The Author(s), under exclusive license to Springer Nature Switzerland AG 2022
G. Boulet et al. (eds.), *Cybersecurity Policy in the EU and South Korea from Consultation to Action*, New Security Challenges, https://doi.org/10.1007/978-3-031-08384-6

China, 2–4, 6–8, 23, 25, 33, 54, 70, 71, 74–76, 81, 82, 84–88, 90, 103, 104, 106, 111, 121, 156, 159, 163, 164, 168, 180, 191, 202, 236–239, 242, 245, 247, 254, 255, 257–259, 262, 266, 270, 275, 276, 279–281
Cloud, 2, 3, 26, 59, 145, 180, 246
Computer, 2, 4, 67, 72–75, 77, 122, 125, 127, 128, 138, 142, 145, 156, 158, 161, 165, 173, 189, 202, 206, 207
Convention, 4, 47, 56, 68, 68n1, 90, 102, 130, 173, 200n19, 202–203, 206–207, 206n44, 207n49, 245, 255, 256, 260, 278, 281, 282
Council of the EU, 46n19, 46n20, 100n4, 105, 106, 147, 147n23, 147n24, 147n25, 147n26, 147n27, 147n28, 148, 148n29, 148n30, 148n31
Countermeasures, 33, 100, 103–105, 109, 109n48, 111, 113, 149, 167, 173, 263
Covid, 5, 18, 24, 42, 45, 46n20, 47, 53, 57, 60, 123, 156, 159, 170, 239, 248, 270, 283
Critical infrastructure, ix, x, 3, 17, 18, 22, 24, 25, 30, 32, 86, 88n69, 112, 145, 147, 163, 166, 180, 241, 270, 275
Cyber, ix–xi, 1–5, 34, 41, 42, 45, 67–92, 99–113, 119–133, 137–151, 155–173, 200, 205, 235, 239–241, 247, 258, 270, 271, 273
Cyber arms, 4, 128, 143, 262
Cyber arms control, 5, 7, 119–133, 139, 140, 144–150, 276
Cyber attack, ix, x, 4, 18, 20–22, 25, 26, 31, 32, 42, 45, 50, 69, 69n3, 71, 72, 75–80, 77n25, 83, 85–87, 89, 91, 92, 100n3, 106–108, 111n52, 112, 120, 123, 124, 126, 127, 131, 132, 139–143, 146–150, 147n23, 147n24, 147n27, 147n28, 156, 158–160, 162, 166, 170, 171, 173, 187, 190, 258, 261, 263, 270, 275–278, 276n17
Cyber capabilities, 70, 77, 82, 83, 85, 86, 89, 90, 99n1, 120–124, 131, 132, 160, 161, 167, 242, 247, 248, 261
Cyberconflict, 70, 71, 122
Cybercrime, 3, 4, 11, 31, 33, 44n8, 45, 45n15, 47, 55–57, 60, 69, 87, 112, 120, 121, 124–126, 132, 144, 146, 158, 164, 169, 173, 182, 190, 198–210, 198n5, 236, 240n7, 245, 247, 248, 255, 256, 261, 265, 276, 278
Cybercrime investigation, 201, 203–204, 208, 210, 215
Cyber Défense, 155–173, 179–193, 254
Cyber diplomacy, xi, 7, 8, 17–34, 46, 105, 142, 148, 185, 237, 240–248, 272, 280
Cyber diplomacy toolbox (CDT), 8, 19–26, 100, 100n5, 105–108, 110n50, 113, 147–149, 147n25, 148n30, 148n31, 149n34, 272, 280
Cyber domain, 8, 21, 68–70, 76, 77, 79–83, 86, 87, 91, 137–151, 245n27, 265, 270, 273, 275
Cyber norms, 67, 88, 88n69, 237, 242, 243n15, 247
Cyber offence-defence, 76–82
Cyber resilience, 10, 21, 24, 55, 78, 143, 144, 170, 179–193, 237, 242–245, 247, 248, 274, 277

Cybersecurity, ix–xxv, 7–10, 12, 17–25, 41–62, 42n2, 68–71, 75–78, 84, 87, 88, 90–92, 105, 108, 111–113, 119, 120, 124, 128, 132, 133, 142–150, 157, 166–173, 168n51, 181–187, 188n37, 189–193, 200, 204, 235–249, 251–267, 270–277, 280–283
Cybersecurity cooperation, 8, 12, 99–113, 132, 241, 242, 248, 251–267, 269–270, 273–284
Cybersecurity strategy, 44n13, 49, 90, 111, 139, 140, 143, 144, 157, 180, 182, 187, 271, 277, 284
Cyberspace, ix, x, 2, 5, 8, 10–12, 18, 19, 21, 22, 24, 27, 28, 30, 41–45, 47, 51, 54, 55, 60, 62, 67, 69, 70, 77, 78, 79n30, 80, 87, 88, 88n68, 88n69, 99n1, 101, 103, 113, 122, 126, 129–132, 139, 140, 142n12, 144, 149, 156, 159, 164, 166–170, 170n57, 172, 173, 180, 181, 187, 190, 199, 202, 202n25, 221n27, 235–240, 240n7, 243–249, 251, 252, 253n1, 254–267, 272, 273, 277, 279, 280
Cyber strategy, 84, 85, 126–127, 165–168
Cyberterror, 120, 124–126, 131, 132
Cyber threat, x, 1–4, 8, 10, 22, 31, 45, 50, 51, 54, 62, 70–76, 83, 88, 89, 91, 100n2, 106n28, 113, 140–142, 149, 156–170, 156n8, 182, 185, 190, 191, 193, 198n4, 205, 210, 235, 241, 244, 246, 247, 264, 265, 270, 272, 275, 280, 281, 284
Cyberwar, 69, 87, 126, 127, 131

D

Data, x, 4, 32–34, 42, 43, 46–48, 47n27, 48n30, 52, 57–59, 58n78, 58n81, 73, 76, 79, 87, 101, 102, 106, 107, 122, 125–127, 138, 147, 150, 179–181, 188, 189, 198n2, 206, 207, 209, 215, 216, 219–224, 226, 228, 236, 238, 244, 244n23, 283
Data protection, xi, 4, 7, 27, 31, 32, 41–62, 146, 168n52, 182, 186, 223, 244n23, 248, 265, 274, 283
Deterrence, 21, 33n63, 46n24, 69, 70, 76, 77, 79, 79n30, 80, 85, 86, 89, 110, 113, 121, 143, 173, 182, 247, 272, 278
Developing countries, 6, 198, 200, 207, 208, 210, 255, 257, 272, 282
Development, 3, 19, 20, 27, 28, 31–33, 50, 58, 59, 77, 78, 89, 121, 122, 124, 129, 130, 132, 141, 143, 144, 146, 150, 159, 162, 180–182, 184, 186–190, 193, 208, 217, 236, 241, 242, 246–248, 253n2, 254, 257, 260, 261, 264, 266, 271–274, 278
Dialogue, 4, 6, 27, 28, 42, 43, 54–56, 57n74, 58, 60, 89, 91, 112, 190, 192, 236, 241, 242, 245, 246, 248, 252, 256, 258, 273, 274
Digital, ix–xi, 1–3, 8, 18, 24, 26, 28–30, 29n52, 41, 42, 46, 49, 52, 54, 60, 62, 70, 72, 75, 77, 91, 101, 125, 138–149, 159, 170, 179–183, 193, 200n19, 201n22, 204, 208, 215, 224, 244, 244n23, 246, 259, 270, 274, 278, 280
Digital forensics, 200, 204, 205, 208–210

Digital governance, 34
Digitalisation, 2, 4, 24, 27, 41, 42, 53, 60, 138, 141, 144, 150, 192, 283
Digital investigations, xi, 7, 216, 219, 226, 228
Digital technology, ix, 2–4, 17–19, 46, 180, 264
Diplomacy, xi, 7, 17–34, 51, 105, 185, 237, 240–248, 274, 280
Diplomacy toolbox, 105–108, 272
Directive, 24, 45, 45n15, 45n16, 47, 48n30, 102, 144, 145, 183, 217n7, 218, 218n8, 219, 224, 225n47, 271
Disinformation, 25, 31, 42, 45, 46, 75, 125, 170, 181, 244, 248, 282
Disruptive, 24, 78, 83, 159, 181, 186–191, 263
DPRK, 156, 158, 160
Dual use, 30, 159

E

East Asia, 70, 71, 76, 81–87, 89–91, 242, 275
Economy, ix, x, 1–3, 6, 8, 17, 18, 26, 27, 29, 34, 42, 49, 54, 56, 60, 121, 122, 141, 159, 170, 180, 192, 244, 246, 247, 259, 270, 274, 283
E-evidence, 101, 102, 203, 215–229, 223n36, 223n37, 223n38, 224n42, 278
Espionage, 3, 4, 18, 68, 82, 85, 87, 91, 92, 126, 157, 162, 164, 165, 245, 275
European Commission (EC), 2, 4, 22, 26, 42n1, 44n13, 46n19, 46n24, 47, 47n27, 48n29, 48n30, 57–59, 58n81, 102, 144n16, 144n17, 148, 150, 150n35, 181–184, 187, 220, 223, 223n36, 223n37, 224, 224n42, 244, 246, 253, 271
European Parliament, 42n1, 46n19, 46n23, 46n24, 47, 47n27, 48n29, 145n19, 150n35, 217n7, 218n9, 224n42, 224n43, 225, 225n47
European Union (EU), ix–xi, 1–8, 17–34, 44–50, 46n23, 46n24, 54–60, 62, 71, 76, 89–92, 100, 105–113, 131, 139, 142–151, 147n23, 147n24, 147n25, 147n26, 147n28, 148n30, 148n31, 155–173, 169n55, 179–193, 198–210, 215–229, 218n9, 218n11, 223n37, 235–237, 240, 243–248, 245n26, 245n27, 252, 253, 255, 258, 264–267, 269–273, 280–284
European Union Agency for Cybersecurity (ENISA), 46n23, 90, 141n7, 145, 145n20, 246, 271
EU-South Korea relations, 251–267

F

Financial, 24n26, 32, 45n16, 58, 72, 91, 108, 119, 125, 158, 162–165, 201, 207, 208
Fireye, 75
5G, 2, 3, 18, 25, 26, 31, 49, 50, 53, 55, 59, 60, 78, 91, 141, 141n7, 170, 184, 244, 246
Foreign affairs, 56, 203, 271
Forensics, 79, 126, 200, 204, 205, 208, 208n52
4th Industrial Revolution, 3

G

Global, x, xi, 2, 3, 6, 19, 24, 26, 28, 30, 31, 34, 41–43, 47–50, 54–56, 59, 71, 78, 87, 101, 111, 121, 130, 132, 133, 146, 150, 151, 156, 157, 159, 163, 164, 166, 169, 173, 180, 186, 190–193, 198, 200, 202, 202n25, 217, 225, 227, 229, 238, 242, 243, 245, 246, 248, 252, 257, 258, 272, 278, 279, 283

H

Hacking, 4, 72, 74, 75, 77n27, 78, 79, 88n69, 119, 124–126, 155, 162, 164, 169, 198, 202, 204
High tech, 3, 18, 71, 192, 206
Honeypot, 89
Hybrid threats, 22, 27, 139, 170, 186, 191, 244, 282

I

Indo-Pacific, ix, 1, 2, 17–34, 169, 274
Information, 3, 18, 23, 43, 46n23, 51, 55, 58, 58n78, 59, 68n1, 71, 72, 74, 75, 85, 89, 90, 101, 106, 107, 119, 121, 125, 132, 138, 139, 141, 142, 145–147, 145n19, 145n20, 149, 150, 158, 160, 165, 166, 173, 182, 183, 185, 186, 189, 191, 199n12, 204, 207, 207n50, 208, 218, 219, 226, 227, 236, 241, 242, 246–248, 253n2, 261, 264, 266, 271, 274, 275, 278, 283
Infrastructure, ix, x, 3, 17, 18, 20, 22, 24, 24n26, 25, 30, 32, 33, 41, 44, 51, 74, 85, 86, 88n69, 107, 112, 121, 125, 139, 141, 145, 147, 150, 163, 166, 170, 180, 183, 184, 241, 252, 259, 265, 270, 275, 276, 284
Intelligence, 21, 32, 72, 74, 78, 80, 82, 85, 88, 91, 92, 100, 107, 127, 155, 156, 161, 162, 170, 186, 188n37, 189, 246, 276
Interdisciplinary, 5
International cooperation, ix, x, 3, 4, 23, 30, 31, 49, 51, 52, 67–92, 112, 157, 166, 167, 173, 193, 200–207, 206n44, 207n49, 210, 216, 218, 228, 235n1, 240–243, 252, 253, 264, 267, 272, 282
International law, xi, 4, 5, 18, 19, 21, 30, 31, 56, 60, 67, 69, 83, 88, 90–92, 102–104, 109, 110n49, 113, 132, 166, 199, 236–238, 240, 246, 248, 253–255, 253n1, 257, 270, 279, 281, 283
Internet, 3, 30, 31, 42, 47, 62, 77, 78, 101, 102, 125, 126, 138, 141, 157, 160, 163, 164, 170, 173, 184, 185, 198, 202, 237, 246, 252, 254, 257, 258, 260, 261, 263, 264, 280
Internet governance, 20, 44, 169, 241, 247, 253n1
Internet of things (IoT), 2, 3, 59, 71, 73, 78, 138, 246, 261

J

Japan, 33, 48, 56, 70, 71, 81, 85, 86, 88, 121, 124, 168, 169, 172, 191, 236, 242, 258, 274, 280, 282

K

Korea, 3, 4, 33, 57, 121, 131–133, 155–173, 244n23, 275, 282

L

Law, xi, 4, 18, 19, 21, 30, 31, 34, 42, 67, 102, 132, 140, 151, 166, 168, 185, 215–229, 236, 253, 270
Law enforcement, 4, 6, 185, 201, 204, 215–229, 240n7, 261

M

Malicious, 7, 18, 20, 21, 31, 32, 46, 71, 72, 78, 80, 99–113, 129, 130, 139, 143, 147–149, 156, 179, 182, 184, 198, 199n9, 208, 242, 247, 270, 276
Member states, 20, 21, 23, 26, 29, 32, 45, 45n15, 47, 49, 55, 76, 105–112, 131, 139, 142, 145, 147, 148, 150, 182, 183, 185–187, 189, 191, 193, 205, 217, 218, 218n8, 220, 221, 223–225, 227, 276, 282
Military, x, 2, 4, 6, 21, 24, 50, 68, 69, 73, 77, 78, 80, 82–86, 100, 107, 108, 121, 122, 126–132, 151, 160–162, 165, 169, 172, 173, 173n66, 187, 189, 241, 242, 245, 256, 259, 263, 264, 266, 276, 277, 279
Multidisciplinary, 186
Multilateral, x, 30, 33, 43, 50, 54, 55, 68, 76, 88–90, 101n8, 112, 133, 167, 170, 173, 218, 228, 235, 240, 242, 243, 247, 251, 255, 258, 272, 275, 280–284
Multilateralism, 4, 5, 30, 31, 270
Mutual legal assistance (MLA), 101, 101n8, 199, 199n12, 209, 216–218, 220–224, 226–229, 278

N

Network and Information Systems (NIS), 145, 185, 246
Norms, 4, 5, 11, 19, 24, 30–32, 34, 47, 54, 56, 60, 62, 67–69, 80, 88, 88n69, 91, 92, 103, 146, 190, 192, 236–242, 246–248, 256, 263, 272, 275, 279–281, 283
North Atlantic Treaty Organization (NATO), x, 21, 22, 44, 90, 112, 129, 131, 190, 191, 235, 246, 252, 254
North Korea, x, 4, 8, 70–73, 75, 76, 84, 85, 87, 88, 100, 106, 119–133, 155–165, 169, 170, 172, 173, 198, 205, 241, 242, 245, 248, 271, 275, 276
Nuclear weapon, 27, 68, 90, 121–123, 159

O

Open-Ended Working Group (OEWG), 56, 88, 88n69, 89, 236, 238, 239, 247, 248, 281

P

Partnership, x, 3, 28, 29, 32, 112, 113, 157, 168–170, 173, 185, 190–193, 237, 243–245, 265, 267
Policy, ix, xi, 2–5, 8–10, 17–34, 45, 46, 49–53, 60, 81, 87, 111, 139, 144, 155, 157, 160, 166, 169–173, 181, 182, 186, 187, 190, 192, 193, 201, 216, 221, 239, 243, 245–247, 249, 252, 255, 256, 258, 263, 265, 267, 270–273, 280

Political, ix, 2, 5, 17, 18, 20, 21, 25, 31, 43, 45, 54, 56, 68, 72, 75, 84, 88, 90, 100, 101n6, 102, 104, 110, 111, 124, 125, 131, 132, 138, 139, 141, 150, 162, 165, 168, 169, 171, 180, 192, 245, 248, 256, 264, 266, 279
Privacy, 3, 31, 32, 34, 43, 48, 57–60, 146, 166, 171, 182, 186, 248, 283
Prosecutor's office, 201, 203, 207
Provider, 2, 11, 33, 101, 141, 145–147, 163, 192, 216, 217, 219–228, 244n23, 278

R

Reconnaissance General Bureau (RGB), 157, 161–163
Republic of Korea (ROK), 10, 57, 58n78, 58n81, 126, 127, 155–173, 192n49, 218n11, 264
Research, 2, 3, 5, 22, 23, 124, 156, 165, 180, 188, 192, 208, 209, 236, 239–241, 246, 247, 253, 261, 274, 283
Restrictive measure, 9, 20, 100, 105–111, 113, 139, 146–149
Retorsion, 9, 99–113
Rights, 4, 11, 18, 19, 27–29, 31, 42, 47–49, 51, 52, 54, 57, 58, 103n17, 107, 166, 170, 184, 203n30, 208, 216, 223–229, 237, 238, 244, 246, 257, 258, 263, 278
Rules, 4, 5, 30, 31, 34, 42, 43, 48, 51, 52, 68, 102, 124, 138, 139, 149, 166, 168, 170, 173, 183, 216, 217n7, 224n43, 225n47, 236, 238, 240, 255–257, 260, 270, 271, 280, 282
Russia, x, 4, 8, 54, 70, 71, 75, 76, 84–86, 88, 100, 103, 104, 106, 111, 129, 131, 141, 156, 159, 163, 164, 168, 236–238, 241, 254, 255, 257, 258, 262, 275, 280

S

Sanction, x, xi, 7, 9, 20, 21, 33, 46, 92, 106, 111n52, 121, 122, 124, 142, 147, 148, 158, 163, 164, 275, 276
Security, ix, x, 3–6, 17–28, 32, 43, 45n15, 45n16, 46, 49–51, 56, 60, 67, 68, 68n1, 73n13, 81–85, 87, 88n69, 89, 91, 108, 110, 112, 121, 128, 131, 132, 139–143, 145, 145n19, 147, 150, 165, 166, 168, 169, 171, 172, 180–184, 186–188, 190–193, 202, 207, 235, 236, 238, 240, 241, 243, 245–248, 259–266, 271–273, 275, 279, 280, 283
Security cooperation, 27, 157, 192, 246, 251, 256, 260, 266, 267, 279
Self-defence, 79, 89, 103n17, 173, 238
South Korea, ix, x, 1–12, 27, 29, 33, 41–62, 70, 71, 76, 81, 82, 85, 88, 90–92, 99–113, 120, 121, 123–127, 131, 132, 155–173, 181, 188n31, 191–193, 191n48, 198–210, 218, 227, 227n50, 235–249, 251–267, 269–284
Sovereignty, 30, 31, 103, 104, 199, 236, 237, 258

State, ix, 2, 4, 8, 18, 20, 21, 23, 26, 32, 41, 45, 45n15, 47, 49, 55, 56, 57n74, 60, 62, 68–72, 75, 76, 81, 82, 85–88, 88n68, 88n69, 90, 92, 101–112, 101n8, 103n17, 104n19, 109n48, 110n49, 119, 121, 128, 130, 131, 139, 141, 142, 145, 147–150, 159, 181–183, 185–187, 189, 191, 193, 202, 205, 207, 216–218, 218n8, 220, 221, 223–228, 236–239, 241, 246, 247, 254, 258, 260, 261, 264, 266, 276, 280–282
Strategic autonomy, 24, 26, 33, 266, 267, 277, 279
Strategy, ix–xi, 9, 11, 19, 44–46, 49, 52, 56, 68, 69, 73, 73n13, 83–85, 99, 111, 112, 119–133, 139, 140, 142–145, 157, 160, 161, 165–172, 180, 181, 205, 210, 260, 261, 263, 270–272, 277, 284
Stuxnet, 73, 77n27, 82, 127, 235

T
Target, 4, 8, 71, 73–77, 73n13, 79, 92, 101, 108, 111, 125, 127, 157, 160, 162, 165, 170, 192, 198n5
Technological sovereignty, 31, 180, 181, 183–185, 187, 191, 193, 272
Technologies, ix, xi, 2–4, 6–8, 10, 11, 17–19, 23–27, 29, 31, 33, 34, 41–62, 68, 69, 73, 77, 83–85, 91, 119–121, 128, 131, 137–139, 141, 142, 145n20, 162, 168n52, 170, 173, 180, 181, 184, 186–193, 199, 199n14, 200, 208–210, 239, 241, 244, 246–248, 253n2, 260, 261, 263, 264, 270, 274, 275, 277, 279, 281, 283
Trade policy, 23

U
UN Group of Governmental Experts (UNGGE), 11, 56, 103, 131–133, 236–238, 253, 253n1, 258, 258n10, 258n11
United Nation (UN), 5–7, 31, 56, 60, 88, 124, 132, 158, 218, 227, 235–237, 239, 240, 252, 253, 253n2, 255, 257, 270, 279–281
United States (US), 2, 6, 8, 28, 32, 48, 49, 70, 71, 73, 75, 76, 81–90, 100, 121, 123, 129, 155, 156, 158–160, 162, 168, 171, 172, 180, 198–200, 204–206, 209, 218, 219, 221–223, 223n37, 236, 237, 240–242, 257–259, 262, 265–267, 270, 275, 279, 280, 282

W
Weapon, 5, 18, 31, 67, 70, 77, 78, 90, 120, 121, 128–131, 138, 142–144, 149, 151, 158–161, 263, 283

SPRINGER NATURE

GPSR Compliance

The European Union's (EU) General Product Safety Regulation (GPSR) is a set of rules that requires consumer products to be safe and our obligations to ensure this.

If you have any concerns about our products, you can contact us on ProductSafety@springernature.com

In case Publisher is established outside the EU, the EU authorized representative is:

Springer Nature Customer Service Center GmbH
Europaplatz 3
69115 Heidelberg, Germany

The manufacturer's authorised representative in the EU is Springer Nature Customer Service Centre GmbH, Europaplatz 3, 69115 Heidelberg, Germany. If you have any concerns regarding our products, please contact ProductSafety@springernature.com

Printed and bound by CPI Group (UK) Ltd, Croydon, CR0 4YY
28/09/2026
02224082-0001